RELATIVITY
AND ITS ROOTS

RELATIVITY
AND ITS ROOTS

BANESH HOFFMANN

DOVER PUBLICATIONS, INC.
Mineola, New York

Photo Credits:

Pages iii, v, vi, xi, 1, 5, 11, 23, 43, 81, 129, 159, The Bettmann Archive

Pages 6, 13, The Granger Collection

Page 20, Reproduced through the courtesy of the Bancroft Library, University of California, Berkeley.

Page 66, Photo Researchers

Copyright

Published in Canada by General Publishing Company, Ltd., 30 Lesmill Road, Don Mills, Toronto, Ontario.
Published in the United Kingdom by Constable and Company, Ltd., 3 The Lanchesters, 162–164 Fulham Palace Road, London W6 9ER.

Bibliographical Note

This Dover edition, first published in 1999, is an unabridged and unaltered republication of the work originally published by Scientific American Books, New York, in 1983.

Library of Congress Cataloging-in-Publication Data

Hoffmann, Banesh, 1906–
    Relativity and its roots / Banesh Hoffmann.
        p.   cm.
    Originally published: New York : Scientific American Books, 1983.
    Includes index.
    ISBN 0-486-40676-8 (pbk.)
    1. Relativity (Physics)—History. I. Title.
QC173.52.H63   1999
530.11—dc21   98-48651
                                                                  CIP

Manufactured in the United States of America
Dover Publications, Inc., 31 East 2nd Street, Mineola, N.Y. 11501

I am grateful to Shaughan Lavine
for many valuable suggestions.

CONTENTS

A Hint of
What Is to Come

CHAPTER 1

A<small>LTHOUGH THIS QUESTION WILL SEEM</small> silly, consider it anyway: Why do the flight attendants on an airplane not serve meals when the air is turbulent but wait until the turbulence has passed?

The reason is obvious. If you tried to drink a cup of coffee during turbulent flight, you would probably spill it all over the place.

The question may well seem utterly inane. But even so, let us not be satisfied with only a partial answer. The question has a second part: Why is it all right for the flight attendants to serve meals when the turbulence has passed?

Again the reason is obvious. When the plane is in smooth flight, we can eat and drink in it as easily as we could if it were at rest on the ground.

Yes, indeed! And *that* is a most remarkable fact of experience. Think of it. In smooth flight a plane can be going at 1000 kilometers (about 600 miles) per hour relative to the ground, and yet inside the plane we do not notice any effect of this uniform velocity.

The same would hold if the plane were any other closed vehicle: There would be no interior effect of its uniform motion. This general statement is of central importance to our story. It is called the principle of relativity, and, as will be seen, it has had a strange history. At this stage there is little, if any, indication of how this principle could be related to nuclear energy, or to the possibility of a man being years older than his twin sister, or to the resolving of a discrepancy between the calculated and observed motions of the planet Mercury. Nor is there any indication of the changes the principle has led to in scientific ideas of time and space. But, as will be seen, the principle lies at the heart of a twofold revolution that stands out as one of the glories of the scientific era.

Do not underrate the importance of time and space. They may seem to be intangible nothings, less palpable even than the faintest breeze.

1

But they are the very stuff of existence; just try to imagine the world without them—or any world at all.

Shakespeare well understood the power and poignancy of time and space. Here, for example, is the song he wrote for Feste the clown in *Twelfth Night* (Act II, Scene III):

> O mistress mine, where are you roaming?
> O, stay and hear; your true love's coming,
> That can sing both high and low;
> Trip no further, pretty sweeting;
> Journeys end in lovers meeting,
> Every wise man's son doth know.
>
> What is love? 'tis not hereafter;
> Present mirth hath present laughter;
> What's to come is still unsure;
> In delay there lies no plenty;
> Then come kiss me, sweet and twenty,
> Youth's a stuff will not endure.

At first this seems artless and lighthearted, as befits the song of a clown. But read it again and note how Shakespeare devotes the first six lines to absence and reunion and thus to imperious space; note also how vividly he devotes the remaining six lines to inexorable time.

The physicist, too, is concerned with space and time. In his work he does not cry out "Journeys end in lovers meeting" or "Youth's a stuff will not endure." Instead, he talks of motion and rest, of distance and duration, and of centimeters and seconds. He measures intangible space and time, and by fitting them quantitatively into equations he makes them sport the formal mathematical garb of his craft. Yet the scientist is no cold automaton. Like the poet, he cannot create without emotion. Behind his equations are audacious imaginings and imperious feelings that transcend logic and give his science an artistry all its own—an artistry that can be made manifest without recourse to detailed mathematics.

Space and time are immensely powerful. In running from physical danger do we not confess the power of space by hoping to use mere distance—space—as a shield? Certainly if space can thus protect us, it is not a soft nothingness.

Nor is time. How safe we would be from death by nuclear bomb had we been born in the time of Shakespeare.

So familiar are time and space that we are apt to take them for granted, forgetting that ideas of time and space are part of the shaky foundation on which is balanced the whole intricate and beautiful structure of scientific theory and philosophical thought. To tamper with those ideas is to send a shudder from one end of the structure to the other. And to effect a profound change in them is to create a major revolution in science and philosophy. In his theories of relativity, Einstein revolutionized our ideas of time and space not once but twice.

Since relativity has roots that reach back to antiquity, Einstein does not officially enter our story until the penultimate chapter. Nevertheless, his ideas haunt these pages. Our tale is that of the historical path to relativity, with its open roads and seeming detours, and Einstein's ideas have been important in determining the attitude here taken toward the past.

In first reading what follows, focus on the overall picture. Details can always be returned to later.

Bon voyage.

The Path
to Newton

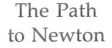

CHAPTER 2

Does the earth move? Primitive people would have unhesitatingly said no. Indeed, they would have wondered how such a question could even be asked. For them a moving earth was unthinkable. The wounded warrior fell to the earth, the stag sped across its surface, and the eagle soared above it. But the earth itself could not fall to the earth like a leaf, or skim its own surface like the wind, or soar above its surface like the sun. Other things could move. But not the earth.

Surrounding the earth as if attesting to the earth's cosmic importance was the awesome vault of the heavens, which seemed to be a sphere set with fixed stars that gleamed like precious jewels, a sphere rotating majestically once a day about the earth. Prominent amid the fixed stars were several wanderers, their total being the mystic number seven: the sun, the moon, and the five starlike planets. (The word "planet" comes from the Greek word meaning "wanderer.") They were called wanderers because, although they shared the daily rotation of the heavens, they moved slowly against the background of the stars.

The sun and the moon were obviously important for humanity. As for the five starlike planets, they came to be named after Roman gods—Mercury, Venus, Mars, Jupiter, Saturn—and, with the fixed stars, were thought to have a major influence on human affairs.

It was natural for mankind at first to think of the earth as fixed with the heavens rotating about it. Much time went by before people of enormous intellectual courage dared to imagine that the earth might move. Two obstacles had to be overcome. The lesser was that everyday experience makes it seem obvious that the earth does not move. The greater obstacle was that a moving earth could not be thought of as the fixed center of the universe, and therefore humanity would be dethroned from its central role in the scheme of things—a frightening conclusion that neither laymen nor theologians, whatever their faiths, would be likely to welcome.

5

The names of the early heroes of the mind who first proposed that the earth might move are probably lost in the mists of prehistory. The first recorded proposal that has survived belongs to the fifth century B.C. It was made by Philolaus, a member of a brotherhood founded by the Greek philosopher Pythagoras, who is probably best known because of the famous theorem concerning right triangles.

PYTHAGORAS PHILOSOPHE
Grec. Chap. 25.

Key developments in relativity will emerge from the Pythagorean theorem. It states that in a right triangle the square on the hypotenuse is equal in area to the sum of the areas of the squares on the other two sides. Thus in the diagram below, the area of the largest square is equal to the sum of the areas of the other two squares.

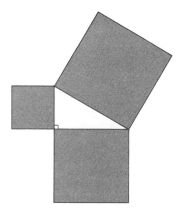

Often, the actual drawing or visualizing of the squares is dispensed with, and the theorem is stated in the powerful but less vivid form that if ABC is a right triangle with the right angle at C, then $AB^2 = AC^2 + BC^2$.

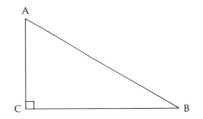

The Pythagoreans correctly believed that the earth is spherical. They believed that the heavenly bodies in their motion give forth musical tones blending into a sublime harmony—a celestial music lost to human ears only because we have been exposed to it every moment of our lives. They looked on numbers as the primary reality and particularly venerated the number 10—the sum of the first four natural numbers 1, 2, 3, and 4, graphically represented by the dots in the mystic triangle below by which they swore their oath of brotherhood. Above all, the Pythagoreans sought beauty in the universe—a theme that will reverberate throughout our story.

According to Philolaus, the earth traced out a circular orbit once a day, always keeping the same face turned toward the center. Nowadays scientists would describe this motion by saying that the earth circled in its orbit once a day and also spun on its axis once a day. The spin accounted neatly for such things as the observed daily rotation of the sphere of the fixed stars, although it is thought that Philolaus ascribed a small motion to this sphere.

To the extent that his system involved both a travelling earth and an earth spinning on its axis once a day, Philolaus had strikingly anticipated modern ideas. But at the center of all things he placed not the sun but a central fire, around which circled not only the earth and the moon but also the sun and the five starlike planets. The circling objects, even counting the sphere of the fixed stars as moving, numbered only nine, and that was an affront to the mystic triangle of ten dots embodied in the sacred talisman of the Pythagorean brotherhood. With beauty and piety alike demanding a tenth moving body, Philolaus introduced a counter-earth that moved so as always to be between the central fire and the circling earth, thus shielding the earth from direct exposure to the fire.

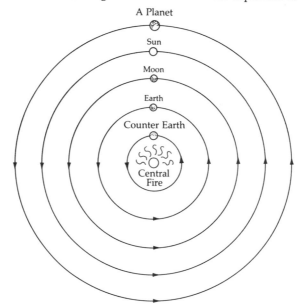

Philolaus' astronomical system

The system of Philolaus came at the dawn of science. For all its quaintness, it deserves our respect. The earth of Philolaus may move in a curious way. *Nevertheless, it moves.*

At an unknown time in the third century B.C., the Greek mathematician Aristarchus of the island of Samos, the island where Pythagoras himself had been born, made a proposal that was even more remarkable than that of Philolaus. What Aristarchus proposed was that not the earth but the sun is at the fixed center of the universe, and that the earth spins on its axis once a day and circles the sun once a year. With such a moving earth there should be perspective changes in the observed positions of the stars, but no such changes were apparent. Nevertheless, Aristarchus did not abandon his idea. Instead he proposed boldly that the stars must be vastly more distant than had hitherto been supposed.

The prophetic concepts put forward by Aristarchus did not evoke any immediate echo of belief. On the contrary, because of his irreverent treatment of the earth, Aristarchus was attacked as impious. There was still powerful resistance to the idea of a moving earth eighteen centuries later, when Copernicus proposed it anew.

This resistance came, in part, from arguments of which the following are two examples. In the fourth century B.C. the Greek philosopher Aristotle argued that objects thrown straight up fall back to the earth at the same place from which they are thrown. If the earth was moving, he asked, would not the thrown objects be left behind during their journeys up and down?

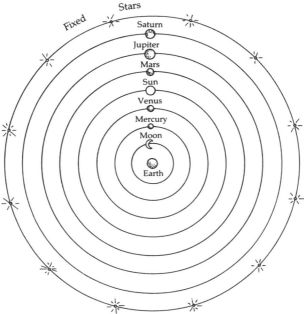

Ptolemy's astronomical system

In the second century A.D. the Alexandrian astronomer Ptolemy argued that if the earth spun on its axis once a day, places on its surface would have speeds of about 2000 kilometers (1200 miles) per hour. Such speeds would create winds and dust storms of unimaginable fury foundering ships, blasting forests, smashing cities, and devastating the face of the earth.

Arguments such as these are indeed persuasive, and to people already disposed to believe in a fixed earth they must have seemed irrefutable. Nowadays scientists respond to such arguments by saying that thrown objects, the atmosphere, and the like are carried along by the moving earth.

Despite Aristarchus, the ancient Greek astronomers continued to place the earth at the fixed center of the universe. In the Greek tradition, the laws governing the heavens were believed to be quite different from those that held sway on the earth—and with good reason, for does not the apple fall vertically from the tree and, on rolling, quickly come to rest even as the moon circles the earth unceasingly?

The work of the ancient Greek astronomers culminated in Ptolemy's masterpiece, known as *The Almagest*, the basic aim of which was to account for the observed motions of the cosmic wanderers. The five starlike planets moved strangely against the background of the fixed stars. Though their main motions were all eastward, each passed through a variety of stages during which it moved westward against the background of the stars. In view of this complexity, the Ptolemaic system is surprisingly simple.

In the heavens, it was believed, there must be eternal perfection, and what more natural and beautiful a representation of eternity could one have than ceaseless motion in that most perfect of figures, the circle? All celestial motions must therefore be circular.

It was a fine ideal, but the facts were against it. The observed motions of the planets could not be accounted for by means of circular orbits around the earth. Remaining as faithful as possible to the ideal of heavenly perfection, the Ptolemaic system accounted for the motions by having the planets move on epicycles—circles whose centers move on other circles.

The agreement with observation was good, and a fixed earth was profoundly satisfying. So, in spite of problems, the Ptolemaic system endured. As century followed century astronomy made no major advances, and the earth remained officially immobile in men's minds. Then, starting in the sixteenth century, a succession of dazzling advances by five great men of widely different temperaments and attainments created not merely a new astronomy but, indeed, a scientific revolution that outshone even the achievements of the Greeks in their heyday. A Pole, a Dane, a German, an Italian, and an Englishman, five towering figures linked by the accident of time and genius, between them ushered in the modern era—in less than two centuries.

The first of the five was Nicolaus Copernicus, who was born in 1473 in the town of Torun, in Poland, and who became a Canon of the Cathedral in Frauenburg, where he now lies buried. Despite his ecclesiastical post, his theological training, and the official belief of the Church of Rome that the earth is at rest at the center of the universe, he dared to propose that not the earth but the sun is fixed at the center of all things, and that the spherical earth spins on its own axis once a day while moving around the sun in a circle once a year. That is just what Aristarchus had said. But Copernicus said it with such cogency and with such a wealth of mathematical detail that at last the idea of a moving earth prevailed—though not in his lifetime.

Aware of the risk involved, Copernicus was reluctant to publish his ideas despite friendly encouragement from high-ranking members of the Church. He did allow a sort of summary to be circulated, but by the time he consented to publish the ideas in full detail, in his famous book *De revolutionibus orbium coelestium*, it was almost too late. A copy rushed from the printer reached him as he lay dying, but he was enfeebled in mind and memory, and it is doubtful that he realized what precious thing had been placed in his hands.

The Copernican system, which is pictured on the following page, had clear advantages over the Ptolemaic. For example, in the Ptolemaic system, with its circles and epicycles, the motion of each of the five planets involved a rotation that takes one year. From the Ptolemaic point of view, that is no more than a fivefold coincidence—an unexplained accident. In the Copernican system, it becomes a fivefold reflection of the annual motion of the earth around the sun, and this is a beautiful simplification. Moreover, the Copernican system allowed the calculation of the relative distances of the different planets from the sun—something quite beyond the capabilities of the Ptolemaic system.

Yet, although this is often not mentioned, there was a strange inconsistency of spirit in the Copernican system: It did not wholly banish the earth from a central role. The unmoving sun was not at the center of

11

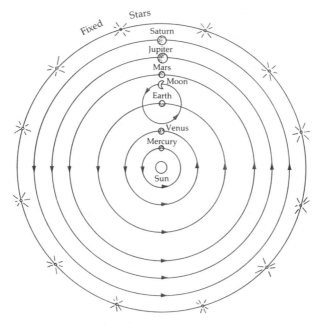

Copernicus' astronomical system

the earth's orbit but somewhat displaced, and the pivotal point of the planetary motions was not the sun but the empty, unsubstantial center of the earth's orbit. Even if less directly than before, the earth was still dominant. Moreover, the Copernican system was not as simple compared with the Ptolemaic as is sometimes thought, since both needed such things as epicycles in order to account for the observed motions of the planets.

A few astronomers were quick converts to the Copernican system. But before it could come to be widely accepted in its detailed form, it was superseded. That fact in itself is a tribute to the enormous influence of the system, since without it the new developments would not have come so soon. Its importance as a turning point in the history of mankind can hardly be exaggerated.

Among those who did not accept the Copernican idea of a moving earth was the Danish astronomer Tycho Brahe, who was born in 1546 and died in 1601. What he objected to in the Copernican system was not its structure but its moving earth. He proposed an alternative that was essentially identical to the Copernican system, except that the earth—instead of the sun—was regarded as being at rest.

Tycho Brahe's objection to the Copernican system and his proposed alternative to it are not what qualifies him as the second of the five

great pioneers who created the modern era. Rather, it is his lifelong dedication to astronomical observation. With the lavish support of royalty he built and operated an astronomical observatory the likes of which the world had never seen. True, it lacked telescopes—they did not exist—but, for its time, Tycho's observatory was a marvel of precision. On his deathbed, in a moment of self-doubt, he cried out in delirium wondering whether his long astronomical labors had been in vain. But by a fortunate set of circumstances Johannes Kepler, the third of the five pioneers, had come to work with Tycho a short time before, and, when death was near, Tycho entrusted to Kepler the fruits of his lifelong labors—the precious records of his observations.

Kepler made outstanding contributions in several branches of science. He was born in 1571 in the town of Weil, Germany, and died in 1630 after a singularly troubled but productive life. He was as much a mystic as he was a scientist. Indeed, his religious mysticism was an essential part of his science. He sought, and often found, beauty and harmony in the heavens. Imbued by the Pythagorean spirit, in one of his books he even gave in detailed musical notation the celestial tunes that he associated with the various planets as they glided in their orbits.

Nowadays we do not believe in those tunes, any more than we

Tunes played by the planets, according to Kepler.

believe in other beauties that Kepler saw in the heavens and for the revelation of which he fervently thanked God. But there are some that we do, indeed, accept.

Of all the planets observed by Tycho Brahe, the most troublesome proved to be Mars. Concentrating on it, Kepler assumed that its orbit was a circle with its center somewhat displaced from the sun and tried to deduce from Tycho's data the size of the orbit and the position of its center. He spent years of ardent, arduous computational labor on the task, and after more than seventy trials found a circular orbit that agreed with the data to within eight minutes of arc—about a quarter of the apparent width of the moon. In those pre-telescope days, most astronomers would have brushed this small discrepancy aside as being simply due to errors of observation. After his laborious calculations Kepler must have been sorely tempted to do just that.

But he could not. He had been close to Tycho and knew the quality of his work. Other observers might make errors of such magnitude, but not Tycho. So Kepler gave thanks to God for having given him this sign of the eight minutes of arc, and with renewed courage continued his search.

Much earlier, following an idea of the English physicist and physician William Gilbert, Kepler had argued that some rotating magnetic influence from the sun must keep the planets moving. But if that were so, the pivot of the planetary orbits ought to be the sun itself, not the empty, unsubstantial center of the earth's orbit, as Copernicus had believed. With this clue, after long calculations of extraordinary ingenuity—in all he spent six years on Mars alone, and his calculations filled

900 pages—Kepler found three laws of planetary motion that are justly celebrated.

Before stating those laws, let me pause to tell briefly about the work of the ancient Greek geometer Apollonius. It forms a background against which to describe not only the laws of Kepler but also a key development that arose from those laws; moreover, it offers an example of the extraordinary ability of mathematicians to make advances that turn out to have quite unexpected applications.

In the third century B.C., in the golden age of Alexandrian mathematics and science, Apollonius studied the curves obtained by slicing what are called right circular cones; not unnaturally, the curves are called conic sections. In addition to circles, they consist of a variety of curves called ellipses, parabolas, and hyperbolas, and they have interesting properties. Take ellipses, for example. They are the shapes of oblique shadows of circles. One can draw an ellipse quite easily by pressing two thumbtacks into a drawing board, placing a slack loop of string around them, and then inserting a pencil in the loop and moving the pencil so as always to keep the loop taut. The thumbtacks are obviously located at specially important points in relation to the ellipse. Each thumbtack is located at what is called a focus of the ellipse.

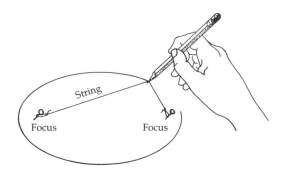

Box 2.1

The important curves called conic sections appear in various sciences ranging from optics to astronomy. Consider a horizontal circle and a point V directly above its center (see top figure page 18). Through V and any point on the circumference of the circle imagine a line to be drawn—not a segment of a line, but a full line of infinite length in both directions. Then as the point moves around the circumference of the circle the line traces out a surface shown. This surface is called a right circular cone, often abbreviated to the single word cone, and the fixed point V is called its vertex.

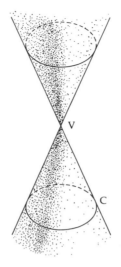

In everyday parlance the word cone conjures up the shape of a dunce's cap or the wafer part of an ice cream cone. But the cone defined above, with its upper and lower parts, counts mathematically as a single cone. Plane slices of such cones are called conic sections.

Since horizontal slices of such cones are circular in shape, circles are conic sections (a). If the slicing plane is tilted a little instead of being horizontal, the shape of the corresponding conic section will be elongated.

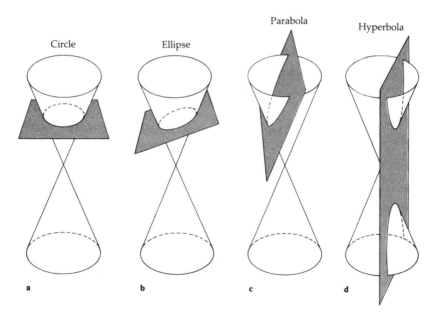

16

One might expect it to be egg shaped, but it turns out to be more symmetrical than that. It is called an ellipse (b). If a tiny light were placed at the vertex of the cone, the ellipse would be an oblique shadow of the original circle. The greater the inclination of the slicing plane, the more elongated the ellipse, until the slope of the plane becomes such that the curve of intersection—the conic section—becomes an open one. It is called a parabola (c). If the slope of the slicing plane increases further, it will cut both the lower and upper parts of the cone. The corresponding conic section will therefore have two separate branches. It is called a hyperbola (d).

The word focus (plural foci) is given to certain special points associated with conic sections. An ellipse (below) has two foci, F and F'. The sum of the lengths FP and F'P is the same for all points P on a given ellipse. Also, if in the plane of the ellipse rays of light were sent out from one focus they would be reflected by the ellipse to pass through the other focus. Strictly speaking a circle has two foci, but they both coincide with its center.

Ellipse

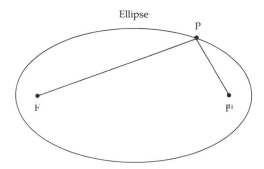

A parabola (below) has only one focus. All rays of light from it are reflected into parallel rays.

Parabola

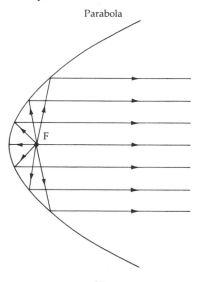

Any given hyperbola has two foci, F and F' such that the difference of the lengths FP and F'P is the same for all points P on it.

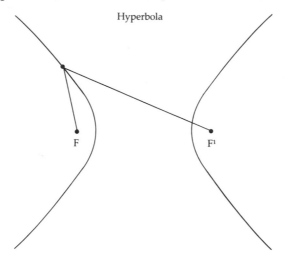

Hyperbola

F   F¹

What led the Greeks to study the conic sections is not known. It could well have been pure intellectual curiosity, but perhaps there was a more practical reason, since the tip of the shadow of a sundial's indicator traces out on a dial a curve that is part of a shadow of a circle.

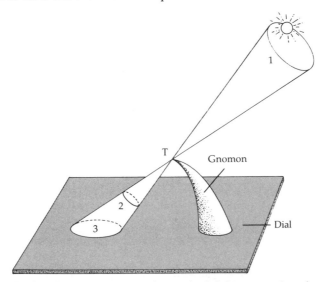

In most latitudes the sun traces out only part of a circle between sunrise and sunset. For simplicity, the diagram uses a complete circle, 1. This circle and the tip T of the gnomon define a cone. A slice such as 2, parallel to circle 1, will be circular. The slice 3 formed by the dial is an oblique one and has the shape of an ellipse.

Here are Kepler's three laws. The first describes the planetary orbits. It says that the planets move in ellipse-shaped orbits with the sun at a focus. The second describes the changing orbital speed of a planet. It says that the line from the sun to a given planet sweeps out equal areas in equal times. The third law is less pictoral than the other two, but no less remarkable. It links the time a planet takes to go once around its orbit to the average of the planet's greatest and least distances from the sun. The law states that if one divides the square of the time by the cube of the average distance, the number obtained is the same for all the planets.

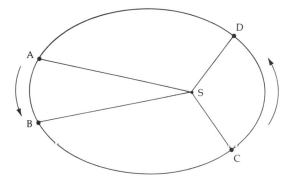

According to Kepler's second law, if the areas SAB and SCD are equal, the planet goes from A to B in the same time it takes to go from C to D. Thus the planet does not maintain a constant speed in its orbit. The closer it gets to the sun, the faster it moves.

The law given first, and usually called Kepler's first law, was actually discovered second. The third law was discovered much later than the other two and was published in the book containing the planetary tunes. The third law may not seem to be linked to the planetary music, but would Kepler have discovered it if he had not dedicated his life to the impassioned search for order and beauty in the heavens?

No one at the time of Apollonius could have foreseen that, some eighteen centuries later, a Kepler would find that the planets move in ellipses. By this and his other laws Kepler transformed astronomy. Gone were the circles and epicycles that cluttered the heavens of Ptolemy and Copernicus. In their place were the lovely shadows of circles, in awesome simplicity.

We come now to the Italian, Galileo Galilei, the fourth of the five great pioneers. He was born in Pisa in 1564, on the very day on which Michelangelo died and in the year that saw the birth of Shakespeare. He was Kepler's senior by seven years, but he outlived Kepler by a decade and more, dying in 1642. The two never met. Both believed that the earth moves, but, strangely, Galileo remained a Copernican rather than becoming a Keplerian.

In telling of the intellectual journey from Philolaus to Kepler, we have been mainly concerned with the motions of celestial objects. Some of Galileo's discoveries and opinions have to do with such motions and are included in this chapter. But his more revolutionary work has to do with the motions of noncelestial bodies. It was this latter work that laid

the foundation of the science of mechanics—the science having to do with forces and their effects on motion. The natural place for presenting Galileo's mechanics is not here but in the next chapter, where it accompanies the masterworks of Newton.

In 1609 Galileo heard of an instrument that made distant objects seem nearer. He immediately devised one himself, and with this telescope and later, more powerful models—all made with his own hands—he explored the heavens. Among other things he discovered four moons circling the planet Jupiter. For him, this discovery was dramatic confirmation that Copernicus had been right, since the moons were circling a body other than the earth—he had found a Copernican system in miniature.

The discovery of Jupiter's moons emboldened Galileo to speak out, and in 1613, in a book on sunspots, he openly championed the ideas of Copernicus. But by now the Church of Rome was becoming concerned about those ideas. In 1616 Pope Pius V officially declared the earth at rest and branded the idea of a fixed sun as heretical. Belatedly for its purposes, the Church placed *De revolutionibus orbium coelestium*, the masterwork of Copernicus, on its Index of Prohibited Books—where the work was to remain until 1822. And Galileo was summoned to Rome and admonished neither to hold nor to defend Copernican ideas.

However, encouraged by the friendly attitude of a later Pope, Urban VIII, Galileo wrote a major book, *Dialogues on the Two Chief Systems of the World, the Ptolemaic and the Copernican,* telling in simple terms of his views of the Copernican system. He wrote not in Latin but in Italian, the language of the people, and he neatly avoided outright advocacy of the Copernican system by using the form of a discussion in which he had three characters look at its pros and cons. He left little doubt, however, as to where his own beliefs lay, and his wit was devastating. The book appeared in 1632. A year later Galileo, aged 70, was summoned by his former friend Pope Urban VIII to appear before the Roman Inquisition, and there, in penitential garb and on bended knee, he was made to swear on the Bible that he abjured, cursed, and detested the error and heresy that the sun is fixed and the earth moves, and that henceforth he would never say or write anything that would cause him to be again suspected of heresy.

He was kept under house arrest, and as part of his punishment he was ordered to recite once every week for three years the seven penitential psalms.

Although confined and shaken in spirit, he was not defeated. Beset by ill health and domestic sorrows, he nevertheless found strength and courage to write a new book, *Discourses on Two New Sciences*—in dialogue form as before. It was his greatest work, presenting the fruits of a lifetime of scientific endeavor. We shall be much concerned with its contents in the next chapter. After considerable difficulty it was published in Leyden, Holland, in 1636. Shortly thereafter, Galileo became blind, but he lived on to the age of 78 and died on January 8, 1642—the very year, it is often remarked, in which Newton was born.

From the psalm Miserère, one of the seven penitential psalms:

> Have mercy upon me, O God,
>   according to thy loving kindness:
> According unto the multitude of thy
>   tender mercies blot out my
>   transgressions.
> Wash me thoroughly from mine iniquity,
> And cleanse me from my sin,
> For I acknowledge my transgressions
> And my sin is ever before me.
> Against thee, only thee, have I sinned
> And done this evil in thy sight.

Newtonian
Relativity

CHAPTER 3

ISAAC NEWTON, THE FIFTH OF OUR FIVE
pioneers, and one of the greatest scientists of all time, was born in the
English hamlet of Woolsthorpe on Christmas day in the year 1642. While
the date is correct as stated, it needs elaboration because of a quirk of
time—albeit a nonrelativistic one. Protestant England was laggard in
going from the Julian calendar to the Gregorian calendar used today.
According to the Gregorian calendar, which was already in use on the
Continent and even in Scotland, Newton was born not in 1642, the year
of Galileo's death, but in 1643, on January 5. In Woolsthorpe, though, he
was indeed born on Christmas day of the year 1642.

In a letter to a scientific colleague, Newton applied an old aphorism
to himself: "If I have seen farther, it is by standing on the shoulders of
giants." He was speaking of his work in optics, but the words apply
more broadly. In this chapter we shall touch on Newton's indebtedness
to Galileo and Kepler even as we tell how his achievements transcended
theirs.

In school, young Newton's early record was hardly auspicious. Indeed, for a while he was at the bottom of his class. Fortunately, the boy
who at that time had the distinction of being just above Isaac Newton in
academic rank kicked him painfully in the stomach—"fortunately," because after winning the ensuing fight, Newton decided to better his
attacker intellectually as well as physically. He succeeded strikingly,
ending his school days as head boy in the school.

He entered Cambridge in 1661. In 1665 the dreaded Black Death
struck London, and soon it spread to Cambridge, causing Newton to
retire for two fateful years to the safety and quiet of Woolsthorpe. There
his genius blazed forth with such intensity that in those two years, in his
early twenties, he laid the basis for almost everything of note that he
was ever to accomplish. He began the construction of the calculus, laid
bare to himself the nature of color, and, he said, discovered the mathe-

matical law governing the amount of gravitational attraction between objects. But he did not hasten to publish his results. Already there was in him a strange tendency toward a secretiveness that was to become almost obsessive after an unpleasant controversy following his early publications on optics.

In 1667 Newton returned to Cambridge, and there, two years later, his mentor Isaac Barrow, who held the newly created Lucasian Professorship of Mathematics, did something truly extraordinary. Recognizing Newton's genius, he resigned so that Newton, at the age of 26, could be appointed to the Professorship in his place.

Years later, in 1684, the English scientist Edmund Halley, who is best known for the comet that bears his name, journeyed to Cambridge to consult with Newton in connection with a scientific argument that had taken place in a London inn. It quickly became clear to Halley that Newton had made extraordinary strides in the study of dynamics and planetary motions. Somehow Halley persuaded Newton to agree to publish his results.

Now the awesome power of his genius took possession of Newton. Barely eating, barely sleeping, barely conscious of his surroundings, he worked with incredible intensity and with unsurpassed insight and technical skill. In a mere eighteen months he completed the main part of the greatest book in the history of science: *Philosophiae naturalis principia mathematica*, now usually called the *Principia*. It was published in 1687.

When the book was completed, he was exhausted and ill. In 1696, being appointed Warden of the Mint, he forsook the cloistered austerity of his Cambridge days to live the far more social life of a celebrity in London, and three years later he was promoted to Master of the Mint, a position he held for the rest of his life.

Honors were showered upon him. In 1703 he was elected president of the Royal Society of London, a post to which he was reelected automatically each year until the end of his days. In 1705 he was knighted by Queen Anne. He died in 1727 at the age of 84 and was buried in Westminster Abbey, and on his grave these words were inscribed in Latin: "Here rests that which was mortal of Isaac Newton."

Some ninety years earlier, the aged Galileo, a virtual prisoner, had dared to write and publish his book of *Discourses*. In it, by challenging long-cherished beliefs, he paved the way for Newton.

For example, practically everyone believed, as Aristotle had, that heavy bodies fall faster than light ones. But Galileo showed otherwise, although he was not the first. It is said that he spectacularly dropped two spheres having widely different weights from the Leaning Tower of Pisa so that people could see for themselves that the spheres kept abreast of each other and struck the ground at the same time. That Galileo used the Leaning Tower is not certain, but he did discover important rules about the motion of falling bodies.

In the *Discourses* Galileo gave a neat argument to show that a heavy stone would not fall faster than a light one. Here it is in a slightly modified form:

Assume that a heavy stone will fall faster than a light one. Then there will be a contradiction, as follows. Consider a stone A and think of it as consisting of two parts, B and C, of equal weight. Since B is lighter than A, it will tend to fall more slowly than A. So, too, will C. And C will tend to fall at the same rate as B. So B and C together will fall at a slower rate than A. But B and C together constitute A. Therefore A will fall at a slower rate than itself, which is impossible.

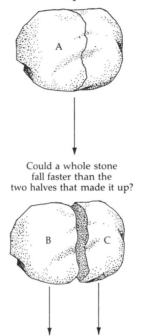

Could a whole stone
fall faster than the
two halves that made it up?

In the *Discourses* Galileo also told of his experimental researches on the problem of falling bodies. Because dropped objects fell too fast for accurate observation of their motion, he "diluted" the gravitational effect by letting spheres roll down gently inclined grooved ramps. He guessed that for a given slope the rate of change of the speed of a sphere with respect to distance would be constant. On persuading himself that this would lead to contradictions, he tried the assumption that he felt was the next simplest, namely, that the rate of change of speed with respect to time instead of distance—what is now called *acceleration*—was constant. And he was able to confirm this assumption by experiment. He began by showing mathematically that if the rate of change of the speed with respect to time, that is, the acceleration, is constant, then the distance traversed from a stationary start will be proportional to the square of the time taken. Thus if a rolling sphere travelled a distance D in time t, it would in time $2t$ travel a distance $(2 \times 2)D = 4D$, in time $3t$ a distance of $(3 \times 3)D = 9D$, and so on.

The main problem in testing the assumption experimentally was to measure time intervals accurately; there were no stopwatches in those days, and a pulsebeat was a poor substitute. Galileo solved the problem by using a large container with a narrow spout at the bottom. He plugged the spout and filled the container with water. For each run of the experiment, he opened the spout at the start and closed it at the finish. By collecting the water from the spout and carefully weighing it for each run, he measured the time taken for that run. True, he measured the time in unusual units, but that did not prevent him from checking that in double the time the sphere rolled four times the distance, in triple the time nine times the distance, and so on. Galileo did more than confirm that, for a given slope, the acceleration is constant. He worked out a theoretical formula linking the accelerations pertaining to different slopes, and, when the formula was confirmed by experiments with various gentle inclines, he boldly assumed it valid for freely falling bodies—which he regarded as analogous to spheres rolling down something vertical, even though freely falling bodies need not spin as they fall. Despite his erroneous neglect of the rotation involved in rolling, Galileo came to the valid conclusion that, if air resistance and certain other factors can be neglected, free bodies dropped near the earth fall with a constant acceleration that is the same for all of them, no matter what they weigh or what they are made of.

Box 3.1

Galileo performed a neat experiment to support the conclusion that the speed of a rolling sphere starting from rest depends solely on the vertical distance traversed. He made a pendulum out of a lead bullet suspended by a fine thread from a nail in the wall at 0. Releasing the bullet from point A, he found that it swung as far as point B, at the same height as A. Next he hammered another nail into the wall, this one at N. When the pendulum was now swung from A, the thread was caught by the nail so that the bullet, instead of following the circular arc CB with center at 0, now followed the circular arc CD with center at N. The end D of its swing turned out to be at the same height as point A. Thus in swinging motion the bullet returned to its original height, and this experimentally supported the conclusion that the corresponding thing would happen in the rolling motion of spheres—whether on straight or curved grooves a rolling sphere would return to the height at which it was released.

Galileo gave another powerful argument in favor of this conclusion. Neglecting air resistance and the like, suppose that a sphere starting from rest at A rolled down AB and up BC and reached C with a nonzero speed. Then by letting it roll along repeated replicas of ABC, such as CDE, one could increase its speed at each stage and thus have a perpetual motion machine, which Galileo regarded as impossible. Thus a sphere starting

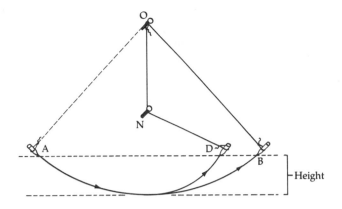

from rest at A could not have nonzero speed at the moment it arrived at C. Suppose, though, that a sphere starting from rest at A did not reach C but reached only as far as X. Then, assuming that leftward motion would be the mirror image of the corresponding rightward motion, Galileo argued that a sphere starting at rest from X would just reach A. Therefore a sphere starting leftward from C, which is higher than X, would acquire a greater speed

in its downward journey and thus arrive at A before its speed was exhausted. So by repeating the grooves as before but in the opposite direction, we would once more have a perpetual motion machine, and this conclusion rules out the possibility that the sphere starting from A would reach only as far as X. Therefore, in the absence of air resistance and the like, a sphere starting at rest at A will reach C when its speed is momentarily zero. The above argument holds for curved as well as straight grooves.

Galileo's discoveries about motion did not stop there. For theoretical and experimental reasons, he concluded that the speed lost or gained by a sphere rolling up or down an inclined groove would depend only on the vertical distance it had travelled. He applied that conclusion— always neglecting friction and the like—to the situation depicted in the diagram on the next page.

Points A and C, D, E, and so on, are all the same height above B, so a sphere rolling down the grooved ramp AB starting from rest at A would acquire just enough speed on its downward journey to let it roll up to the top of any of the other ramps level with A. The ramps BC, BD, and so forth, are successively longer, and the sphere would therefore roll farther on each ramp than it did on the ones before.

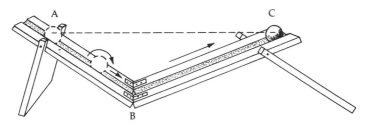

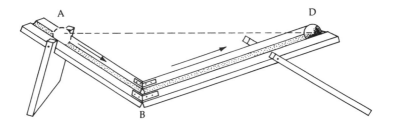

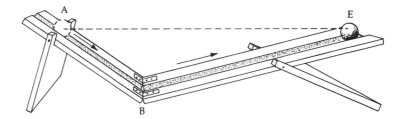

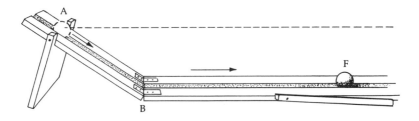

What if a ramp were made horizontal? Then, since that ramp would never reach the height of A, a staggering thing would happen: The sphere would continue rolling forever, and indeed, since its height would remain unchanged, its speed, too, would remain unchanged.

In practice, of course, the sphere would come to a stop. But Galileo realized that this was due to friction and air resistance, which were obscuring the basic law. He concluded that the natural motion of a free particle is uniform motion in a straight line. The French philosopher René Descartes came to a similar conclusion on quite different grounds.

The discovery of this law revolutionized the science of mechanics. To sense the law's importance, all we need do is think for a moment of everyday experience. It tells us, as it told Aristotle and his many influential followers up to the time of Galileo and even beyond, that the natural thing for a moving object to do if left alone is to come to rest. If we see something moving uniformly, we are likely to ask naively what force is keeping it moving. It all seems obvious. But what of an archer's arrow? That or any other projectile seems to pose a problem. What keeps the arrow moving once it has left the bow? The followers of Aristotle had an ingenious answer: The initial thrust of the bowstring continues to be transmitted to the arrow by the air.

Box 3.2

René Descartes approached problems philosophically rather than experimentally. He believed that motion in a straight line is simpler than motion in a circle. Here are excerpts from his book Principia Philosophia:

> As regards the general cause [of motion], it seems clear to me that it can be none other than God himself. He created matter along with motion and rest in the beginning; and now, merely by his ordinary cooperation he preserves just the quantity of motion and rest in the material world that he put there in the beginning. ... Further, we conceive it as belonging to God's perfection not only that he should in himself be unchangeable but also that his operation should occur in a supremely constant and unchangeable manner. Therefore, apart from the changes ... that we can see, or believe by faith, that they take place without any change in the Creator, we must not assume any others in the works of God, lest they should afford an argument for his being inconstant. Consequently it is most reasonable to hold that, from the mere fact that God gave pieces of matter various movements at their first creation, and that he now preserves all this matter in being in the same way as he first created it, he must likewise always preserve in it the same quantity of motion.

According to Galileo, scientists had been asking the wrong questions. If an object kept moving uniformly in a straight line, there was no need to ask what was keeping it moving, no need to offer an explanation or an apology. On the contrary, an explanation was called for if the body slowed down and came to rest; some force, probably friction, must have caused it thus to depart from its natural uniform motion in a straight line.

There is an element of irony in the manner by which Galileo came to his discovery. Recall that he decided that in the absence of air resistance and the like, a free sphere rolling on a horizontal groove would do so with uniform speed forever. But "horizontal" meant of constant height, and on a spherical earth a line of constant height would not be straight and of infinite length; it would gird the earth and thus be circular. Assuming, as usual, that the "horizontal" groove counteracted the pull of the earth's gravity, one could argue, therefore, that what Galileo had really shown was that the natural tendency of a free particle is not to move uniformly in a straight line but to move uniformly in a circle concentric with the earth—a decidedly pre-Copernican concept and one seeming to confirm the ancient idea of the celestial perfection of circular motion around the earth. Galileo was not unaware of the irony.

Among other things, Galileo applied his discoveries to the motion of cannonballs—a motion that had never before been properly understood; learned people had even believed that objects hurtling through the air continued until their speed was exhausted, and then fell vertically to the ground. With his new insights, Galileo was able to solve the problem of the motion of cannonballs.

He argued that a cannonball would combine a horizontal motion and a vertical motion, each of which he had already investigated. The horizontal motion would be uniform, but the vertical motion would be with constant downward acceleration. When combined, these resulted in a trajectory having the shape of an arc of a parabola—one of the conic sections.

Box 3.3

Galileo gave the first acceptable theory of the motion of cannonballs. For simplicity, take the case of a cannonball shot horizontally. Galileo argued that its motion would combine a horizontal motion and a vertical motion, each of which he had already studied. Horizontally, the cannonball would move at constant speed, like a sphere rolling on a horizontal groove. Vertically, it would fall with constant acceleration, keeping pace with a dropped object.

Successive dots mark positions at the ends of successive seconds. Note that the cannonball shot horizontally from the cannon keeps pace vertically with the one dropped from the same starting point and keeps pace horizontally with the one rolling uniformly on a horizontal groove. Since vertical

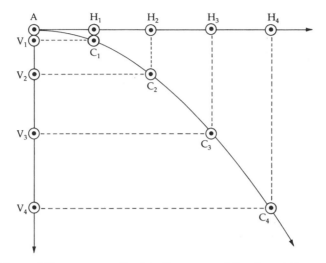

distances fallen are proportional to the square of the elapsed time and horizontal distances traversed at uniform speed are proportional to the elapsed time, the vertical distances are proportional to the squares of the horizontal distances, a relation that happens to be characteristic of parabolas.

Let us now see what happened when Newton, following important hints from his colleague Robert Hooke, applied his powerful intellect to the question of the motion of cannonballs. Assume for simplicity that air resistance can be neglected and that there are no objects in the cannonball's path. Assume also that the cannon is vastly more powerful than any actual cannon, and that it moves out of the way after each shot. Fire the cannon horizontally from the top of a high tower. Then the greater the initial speed of the cannonball, the farther away it will reach the ground. But the earth is round; its surface curls away from the cannonball. If the cannon were sufficiently powerful, it could shoot the cannonball halfway around the earth. Given yet higher muzzle speeds, the cannonball would curl farther and farther around the earth, landing closer and closer to the tower. With enough initial speed, it could even go all the way around the earth and strike the foot of the tower. And with yet a little more initial speed, the cannonball could miss the earth completely and arrive at the top of the tower moving horizontally, in which case it would repeat its earth-circling motion, never landing even though it is continually accelerated toward the spherical earth. The cannonball would become what is nowadays called an artificial satellite—a circling body akin to the moon.

Box 3.4

Newton went beyond Galileo's theory of the motion of cannonballs. The diagram shows trajectories of cannonballs shot horizontally at various speeds. If the earth is assumed to be flat, the cannonballs all reach the ground at the same time. But the earth is not flat. The corresponding diagram for a spherical earth is shown next. The surface of the earth keeps curling away from the falling cannonball, and the "vertical" pull of the earth's gravity keeps changing direction. The two effects tend to cancel each other, but not completely. If the initial speed is great enough, the cannonball misses the earth while "falling" around it—an artificial satellite.

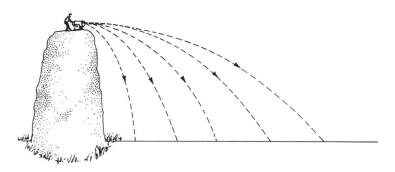

It may seem strange to speak of the circling cannonball as falling when it is maintaining a constant distance from the earth. But scientists think of velocity as having direction as well as speed. An object moving in a circle with constant speed is continually changing its direction of motion and thus also its velocity, despite its unchanging speed. Since its velocity changes it is, by definition, accelerated. For circular motion with constant speed, this acceleration turns out to be directed toward the center. In this sense we can see from a new point of view how the circling artificial satellite is continually falling toward the earth.

Consider a point initially at P and moving in a circle, center O, with constant speed. Velocity involves both speed and direction. If P were unaccelerated—that is, if its velocity were constant—its motion at P would be with constant speed along the straight line PT that is tangent to the circle at P. Clearly, OT is greater than OP and OQ. If this unaccelerated motion brings an increase in radial distance, then constant radial distance must involve acceleration toward O, even though the circling speed does not change. The radial effect is acceleration: Even for a constant speed, a changing direction of motion implies a changing velocity and thus acceleration.

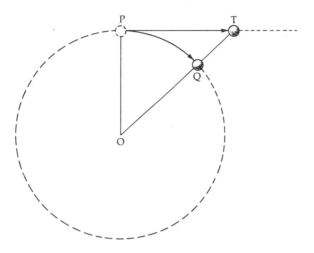

The story goes that during the plague years 1665–1666, as young Newton sat in the quiet of his garden in Woolsthorpe, the fall of an apple set him to wondering. The force of gravity that drew the apple to the earth certainly extended to heights far greater than the height of the apple tree. It was present even atop high mountains, and surely it did not suddenly cease there. What if it reached as far as the moon? Then it might make the circling moon and the falling apple fellow captives of the earth. And a similar gravitational force from the sun might hold its planetary flock in thrall.

How could Newton test his idea? By comparing apple and moon. For if both were tied to the earth by its gravity, their accelerations toward it should be linked. First he must know if gravitation weakened with distance, for the pull of the earth might be much enfeebled at a place as remote as the moon.

Here Kepler's third law was the key. From it, assuming a circular orbit for simplicity, Newton found that the gravitational force falls off as the inverse square of the distance. This means that if the distance is doubled, the force is reduced to $\frac{1}{2^2}$, or $\frac{1}{4}$, of what it was; if the original distance is tripled, the force is $\frac{1}{3}^2$, or $\frac{1}{9}$, of what it was; at four times the distance, the force is $\frac{1}{4}^2$, or $\frac{1}{16}$, of what it was; and so on. Armed with his inverse-square law, Newton could now make the crucial test by which his whole speculative structure would stand or fall. Knowing the acceleration with which an apple fell and knowing the law by which the pull of gravity was enfeebled by distance, he could calculate the acceleration with which the moon should be falling if it were being held by earth's gravity. But he could also calculate that acceleration directly from the distance of the moon from the earth, and from the fact that it circled the earth once a month. The calculations had their shaky aspects. For example, young Newton had to guess that the distances of the apple and the moon should be those from the center of the earth—a theorem that he was able to prove only much later. How did the two numerical values for the acceleration compare? According to Newton, he found them to "answer pretty nearly," this phrase coming from a recollection that he set down some fifty years later.

Box 3.5

Here is essentially the way Newton must first have derived the inverse-square law of gravitation. For simplicity, he took the case of a circular orbit. Also, he assumed that the gravitational attraction of a uniform sphere was the same as it would be if all its mass was concentrated at its center—a result that he was apparently unable to prove until much later.

Consider a planet in circular orbit around the sun. Denote by R the radius of the orbit, and by T the period of the orbit—the time needed for one circuit of the orbit. Using the symbol PROP to stand for "is proportional to," one may write Kepler's third law in the form

$$T^2 \text{ PROP } R^3. \tag{1}$$

The orbital speed, v, is proportional to R/T, so

$$v \text{ PROP } R/T. \tag{2}$$

As Newton's contemporary Christiaan Huygens had already found, and as Newton discovered independently, the rate of fall of the planet toward the sun is proportional to v^2/R. Newton argued that this acceleration toward the sun must be proportional to the gravitational force acting on the planet from the sun. Denote this force by F. Then,

$$F \text{ PROP } v^2/R,$$

so, by (2)

$$F \text{ PROP } R/T^2.$$

From this, by (1)

$$F \text{ PROP } R/R^3,$$

and thus

$$F \text{ PROP } 1/R^2.$$

If the numerical results did "answer pretty nearly," that was surely a stupendous fact. But what did Newton do? Apparently he kept this magnificent discovery to himself and turned to the study of optics.

Because of this strange delay, and a scientific error that Newton made in a letter to Hooke as late as 1679, doubts have been expressed that he made the above discoveries about gravitation as early as he said. But there is no doubt at all about the overwhelming greatness of his *Principia*. In it, in five short sentences, he enunciated three laws of motion, and with those and his law of universal gravitation he revealed a breathtaking unity of the heavens and the earth: The same physical laws hold sway throughout the universe.

His first law of motion, often called the "law of inertia," says that every body continues in a state of rest or of uniform motion in a straight line unless it is compelled to change that state by forces acting on it. This law is far more than a mere restatement of Galileo's discovery—not in its wording, but in its setting. For Galileo tended to be earthbound, but Newton sought a grand synthesis of the heavens and the earth and dared to give his laws cosmic validity.

To appreciate the depth of Newton's theory, imagine that we are out in space, far from the earth. We want to make sure that a body is moving in a straight line. For simplicity, let the body be a bead threaded on a wire. If the wire is straight and the bead is moving along it, then surely the bead is moving in a straight line.

But no—and this is a crucial point. Suppose someone is whirling the wire as if it were the baton of a drum majorette. Then, in spite of the straightness of the wire, we would not say that the bead was moving in a straight line. However, we would certainly be ready to say so if, for example, the straight wire was kept at rest. But how do we keep it at rest? There are no stationary milestones or other markers cemented into space against which to recognize rest. Returning to the earth and clamping the wire in sturdy laboratory vises would not help. Nor would letting spheres roll on straight grooved ramps. According to Copernicus, the earth is not at rest. It is pirouetting on its axis and orbiting the sun.

How, then, can we give cosmic sense to the ideas of rest and motion in a straight line, now that we no longer have a fixed earth? There is no solution. But Newton needed one, so he invented one. His solution may seem simple enough today because it has been a part of our culture for centuries. He envisaged a universal absolute space that was, by fiat, forever and everywhere the same and forever and everywhere at rest: "Absolute space in its own nature, without relation to anything external, remains always similar and immovable." Although Newton's absolute space was utterly featureless, it allowed him at least to talk of a wire or any other body as being at rest in it or as being in motion relative to it—and thus as being at absolute rest or as being in absolute motion.

So much for rest and for motion in a straight line. But how can one tell if a straight-line motion is uniform? One direct way would be to mark off equal lengths along the line and note whether they are traversed in equal times. But measuring the times requires a clock, and if the clock is erratic it will make uniform motion seem nonuniform. How can one tell whether a clock is trustworthy? Against what standard can it be tested? And how can one know that this standard is itself free of error? There are really no satisfactory answers to such questions. But the questions themselves do reveal that in the back of our minds a belief in a "true" time does exist. Behind the discrepancies of clocks we dimly conceive of Time itself, an invisible majesty, undefinable yet apprehended by us all. Newton, therefore, envisaged absolute time, or, to use his full phrase, "absolute, true, and mathematical time," saying that it "of itself, and from its own nature, flows equably without relation to anything external, and by another name is called duration."

That absolute time flows uniformly is a tautology. For how could the uniform flow of absolute time be tested except against absolute time itself, and in that case how could its flow turn out to be other than uniform?

I by no means intend to belittle Newton's concepts of absolute space and absolute time. They were dazzlingly effective inventions, and it was part of Newton's genius that he dared to build his theory upon them in spite of their problems.

With absolute space and absolute time, Newton could express his laws in a cosmic setting. We have already discussed the first law. We see it now in a new light. It tells us that in the absence of force a particle will either be permanently at absolute rest in absolute space or else be moving along a straight line fixed in absolute space, and doing so with a speed that is constant according to absolute time.

So much—but only for the time being—for Newton's first law of motion. His second law tells the effect of force on the motion of a particle. It is often summed up in the phrase, "force equals mass times acceleration."

$$F = ma$$

From this law it follows that, for any given force, the larger the mass of the body on which it acts, the smaller the acceleration that the force gives rise to, and vice versa. This consequence of the law accords with everyday experience: the more massive an object, the harder it is to budge it or, if it is already in motion, to slow it down or speed it up. Thus the mass of a body is a measure of its, shall we say, "reluctance" to be accelerated by forces. The technical term for this reluctance is inertia.

Newton's third law of motion says that if one body exerts a force on another, the second exerts an equal but opposite force on the first. This law can seem quite incredible, since, according to it, the gravitational pull of the earth on the apple and the gravitational pull of the apple on the earth are of equal magnitude. But Newton had made typically neat experiments attesting to the law's validity.

For example, on the surface of some water he placed three floats. On one float he placed a magnet and on another he placed a piece of iron. On the third float he placed a partition between the other two. If the force exerted by the magnet were the stronger, the whole assembly would be pulled toward the magnet; if the force exerted by the iron were the stronger, the assembly would be pulled toward the iron. Since the system stayed at rest in the water, he concluded that the force with which the magnet pulled the iron balanced the force with which the iron pulled the magnet.

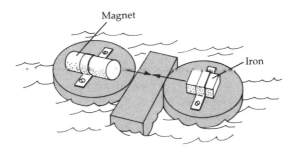

Let us now briefly note the magnificent structure that Newton erected on the meager and seemingly inadequate foundation of his three laws of motion. From Kepler's law of equal areas swept out in equal times he showed that the gravitational force between the sun and a planet is along the line joining their centers. (No transverse force was needed to keep the planets moving; inertia was sufficient.)

We have already seen how, by using Kepler's third law and assuming circular orbits, Newton may have concluded that the gravitational force diminishes as the inverse square of the distance. In his *Principia* he showed something much stronger. Suppose, as is almost the case, that the masses of the planets are negligible compared with the mass of the sun. Then from Kepler's discovery that planets move in ellipses with the sun at a focus, Newton proved that the gravitational force would have to diminish as the inverse square of the distance. He showed, moreover, that the same conclusion would hold if the orbit had been any conic section—ellipse, parabola, or hyperbola—with the sun at a focus. And there was something stronger yet: If the gravitational attraction diminishes as the inverse square of the distance from the sun, then the planetary orbits cannot be other than conic sections with the sun at a focus. (Recently Robert Weinstock has drawn attention to a gap in Newton's proof of this result. The theorem is valid however.)

All these discoveries are among the riches in the *Principia*. What is strange is that nowhere in the book does Newton fully state his law of universal gravitation prominently and all in one place. (The fullest statement occurs almost as an aside in corollary 4 of Proposition LXXVI in Book I.) Let us remedy this lapse by gathering the pieces together. According to Newton, every particle in the universe attracts every other particle in the universe by an instantaneous attraction at a distance. If one particle has mass m and another has mass M, and if the distance between them is r, their mutual gravitational attractions are proportional to the quantity mM/r^2, and these attractions are radial, that is, they are exerted along the line joining the particles.

Later we shall be particularly interested in the way Newton's theory accounts for Galileo's law of falling bodies, the law that states that, neglecting air resistance, all bodies fall with the same acceleration, no matter what their masses and no matter what they are made of. Newton gave mass a double role to play. It had a gravitational effect, and it also measured the inertia of a body. Compare the behavior of two falling bodies, one having twice the mass of the other. Doubling the mass does two things: It doubles the gravitational pull of the earth on the body, and at the same time it doubles the body's inertia, its reluctance to be accelerated by that pull. As a result, the second body has the same acceleration as the first. Obviously the conclusion holds for bodies of all masses.

Another way of talking about how Newton's theory accounts for Galileo's law is to use the terms inertial mass and gravitational mass. The purpose is to make a sharp distinction between the different roles of mass. Inertial mass refers to the mass of a body considered solely as a measure of its reluctance to be accelerated; gravitational mass refers to the mass of a body considered solely as a measure of its gravitational effects. Thus inertial mass and gravitational mass present themselves as quite different concepts; however, in Newton's theory, Galileo's law of falling bodies follows from inertial and gravitational masses both being measured by the single quantity that Newton simply called mass; that is, Galileo's law follows from the equality of gravitational and inertial mass.

Newton's theory applied equally to all bodies, whether in orbit in the heavens or dropped from towers or shot from cannons on the earth. By it people would predict the times of return of comets, locate unknown planets by their gravitational effects on the motions of known ones, and even send men to the moon and instruments to Mars and beyond in search of life. It was stupendous.

But there were problems. One was that, much as he tried, Newton could offer no plausible mechanism that would account for the inverse-square law of gravitation. The gravitational force seemed to act instantaneously at a distance. Newton himself said:

> That gravity should be innate, inherent
> and essential to matter, so that one body
> may act upon another at a distance
> through a vacuum, without the mediation
> of any thing else, by and through which
> their action and force may be conveyed
> from one to another, is to me so great an
> absurdity, that I believe no man who has
> in philosophical matters a competent
> faculty of thinking, can ever fall into it.

A second problem concerned absolute space. In his *Principia* Newton argued powerfully for the absoluteness of space by describing a beautifully simple experiment involving rotation. He suspended a pail by a long cord and then twisted the cord tightly. Next he poured water into the pail, which he then allowed to spin, starting it off with a sudden twist. The cord unwound, sustaining the pail's rotation. At first the surface of the water was flat, but as the rotation of the pail communicated itself to the water, the surface became concave, and the concavity increased until the water was rotating at the same rate as the pail.

The experiment itself, essentially as Newton described it, seems inconsequential. But Newton knew what he was about. He asked what caused the concavity of the surface of the water. It could not be the rotation of the water relative to the pail, since there had been no concavity at all at the start, when the difference in the rates of rotation of pail and water was greatest; and the concavity was actually greatest when the relative rate of rotation of pail and water was zero. Thus the concavity of the surface of the water must be revealing the water's absolute rotation—and that, said Newton, could only be with respect to absolute space.

Important contemporaries of Newton's, notably the British philosopher Bishop George Berkeley and the German diplomat, philosopher, and mathematician Gottfried Wilhelm Leibniz, were highly critical of Newton's absolute space and time. Indeed, Leibniz argued that space and time have no independent existence but are merely relations among pieces of matter. In the nineteenth century, philosophical mistrust of the concepts of absolute space and time was re-echoed and amplified by the Austrian philosopher and physicist Ernst Mach. He objected in part because absolute space conflicted with Newton's own third law of motion. According to that law, if one object exerts a force on another, the second will exert an opposite force on the first. But although, in Newton's theory, the inertia of a body was a measure of its resistance to being accelerated relative to absolute space and arose therefore from an action of absolute space on the body, there was no opposite action of the body on absolute space; for by definition absolute space was utterly unaffected by what went on within it.

Rejecting the concept of absolute space, and with it absolute motion, Mach argued that the inertia of a particle must arise from some interaction between it and all other matter in the universe, particularly the more distant matter; he often spoke of this distant matter as the fixed stars. According to Mach, the concavity did not arise from absolute rotation of the water relative to absolute space. It arose instead from rotation of the water relative to all the other masses in the universe. Therefore, argued Mach, one should obtain the same concavity if the water was "not rotating" but if, instead, all the rest of the universe were rotating about it with the same relative rate of rotation as before. Mach's ideas, expressed philosophically rather than in specific mathematical form, were to have a considerable influence on Einstein.

Inevitably, relating the criticisms of the foundations of Newton's theory tends to belittle his achievements. Einstein, in his *Autobiographical Notes*, came face to face with this problem and solved it with a graciousness no one else was in a position to match. Having discussed objections to Newton's theory as a necessary preliminary to a presentation of his own general theory of relativity, which went beyond Newton's, Einstein broke off abruptly and addressed Newton directly across the centuries with these words:

> Enough of this. Newton, forgive me. You found the only way that, in your day, was at all possible for a man of the highest powers of intellect and creativity. The concepts that you created still dominate the way we think in physics although we now know that they must be replaced by others farther removed from the sphere of immediate experience if we want to try for a more profound understanding of the way things are interrelated.

The criticisms by Berkeley and Leibniz led Newton, in his seventies, to insert a special section—the famous "General Scholium"—at the end of his *Principia*. Here are excerpts from it:

> The Supreme God is a Being eternal, infinite, absolutely perfect.... He is eternal and infinite, omnipotent and omniscient; that is, his duration reaches from eternity to eternity; his presence from infinity to infinity; he governs all things and knows all things that are or can be done. He is not eternity and infinity, but eternal and infinite; he is not duration or space, but he endures and is present. He endures forever, and is everywhere present; and by existing always and everywhere, he constitutes duration and space.... God suffers nothing from the motion of bodies; bodies find no resistance from the omnipresence of God.

And in a book on optics Newton spoke of absolute space as the sensorium of God.

Particularly disturbing to Newton was a result that he deduced quite easily from his laws. He gave it no name, but it is now conveniently referred to as the Newtonian principle of relativity. It will remind us of the airplane in smooth flight. Here, in a standard but somewhat quaint translation of the Latin of the *Principia*, is Newton's statement of this principle of relativity: "The motions of bodies included in a given space are the same among themselves, whether that space is at rest, or moves uniformly forwards in a right line without circular motion." The word "right" means straight. Here the word "space" does not refer to absolute space but to the space inside a vehicle. Put another way, Newton's statement says that inside a laboratory moving uniformly in a straight line in absolute space and not rotating, no mechanical experiment will reveal the laboratory's motion—all mechanical processes inside the laboratory will take place as if the laboratory were at rest.

It is not hard to see why the principle of relativity would be disturbing to Newton. By introducing absolute space, he had provided a setting in which one could make a sharp distinction between rest and motion. But in practice—according to Newton's own laws of motion—there was no actual, physical, observable distinction between rest and uniform motion, that is, motion in a straight line without rotation, although there still was a distinction between rest and nonuniform motion (recall the airplane in turbulent flight). According to Newton's laws, rest and uniform motion are relative, which is to say, not absolute; they belie the absoluteness bestowed on them by absolute space and absolute time. In principle rest and uniform motion are absolute; in practice they are not.

Newton approached the problem in a subtle way. He knew that although the masses of the planets were small compared to the mass of the sun, they could not wholly be ignored. He pointed out that since the sun attracted the planets, the planets would attract the sun, pulling it this way and that. Therefore, as Newton put it, the sun would be "agitated by a continual motion" that, although small, was far from uniform. Ignoring the stars he showed, however, that there was one spot in the solar system—its center of gravity—that was unaccelerated. Most of the time, this unique spot lay inside the sun, and it never strayed far outside. Being unaccelerated, it was either at rest or in uniform motion. Next Newton added to his system of laws: *Hypothesis I: That the center of the system of the world is immovable.*

To justify this he remarked that everyone believed it, though some would place the sun at the center, and others the earth. But Newton knew that the sun and the earth, being both accelerated, could not be at rest. For him the only candidate for the role of the center of the world was the center of gravity of the solar system, which must be either at rest or in uniform motion. Declaring it to be the center of the world, he pointed to his Hypothesis I as showing that it must be at rest.

This maneuver established a fixed point, and that, it turned out, was sufficient to establish absolute rest and absolute motion everywhere. But Newton had had to bring in a specially tailored hypothesis from outside his laws, and his having done so is an indication of his unease about the principle of relativity implied by those laws.

One might well be expecting now to see modern relativity emerge as an immediate offshoot of Newton's unease. But the story of relativity is more surprising than one would likely imagine at this stage. We have to take what may seem like a detour on the road to relativity by exploring highlights of the theory of optics and the theory of electricity and magnetism. This we do in the following chapter, and it will lead us directly to relativity.

The Optical Threat
to Relativity

CHAPTER 4

DOES LIGHT MOVE? IN PREHISTORIC TIMES, people would have completely misunderstood the question. They probably would have pointed to the waving patches of sunlight under a swaying tree and remarked that, of course, light moved. For them—as for children from time immemorial—darkness, not light, was the greater presence. The sun, the moon, and the stars took turns to do battle with darkness. But darkness could bide its time. Some day the sun and the moon and the stars, like fires on the earth, might give up the struggle, and then darkness would be the victor. For even when all was gone, would not darkness still remain?

Imagine the boldness of those persons who first dared to conceive of light as a presence and darkness as mere nothing, and the subtlety of those who then conceived of a physical light that is neither a source of light nor a sensation of light but an agent linking the two. It is about this agent light that we ask whether there is movement. Is it a journeying thing? Or something instantaneous that takes no time to bridge the gap? For a long time people believed the latter—even Kepler did. Galileo seems to have been the first to put the matter to experimental test. He tells about it in his *Discourses*, which he wrote in the bitter late years of his life when he was under house arrest.

Galileo had two men face each other about a kilometer and a half (less than a mile) apart on hilltops. Each carried a lantern, which he hid with his hand from the other. One of them abruptly lifted his hand so that light from his lantern could travel to the other man. As soon as the second man saw this light, he lifted his own hand so that light could travel from his lantern back to the first man, who noted the time that elapsed from the moment when he lifted his own hand until he saw the

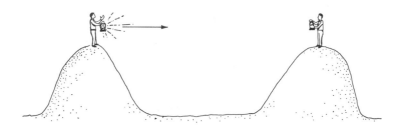

return signal. This elapsed time would be the time taken for light to travel from one man to the other and back; from this information one could calculate the speed of light.

We may smile as we think of the time it took the second man to react to the light from the first, and for the first to react to the light from the second. They could not have reacted instantaneously—after all, people are only human. But let us not underrate Galileo. He was well aware of the problem and took steps to overcome it. He had the men rehearse the routine at close quarters until they were adept and noted how long they took, so that when they went to the hilltops he knew how much of the total delay was due to the men and how much to the journeying of the light. He found that all the delay was caused by the men.

From the results of the experiment Galileo concluded that light might well be instantaneous, but if not it must certainly be extraordinarily rapid. (Actually, to travel 3 kilometers, or 2 miles, light would take less than a hundred-thousandth of a second.) One of Galileo's remarks is charmingly revealing. He suggests that the experiment be repeated with the men separated by twice or three times the distance, saying that if no effect were discernible with a total distance of 9 kilometers (6 miles) there and back, one could safely conclude that light is instantaneous.

We recall that Galileo discovered four moons circling the planet Jupiter. There may seem to be no connection between this fact and the speed of light. But science often takes unexpected twists. The moons of Jupiter, like our own moon, shine not by their own light but by reflecting the light of the sun. When Jupiter stands in the way of that light its moons are eclipsed. In Paris in the early 1670s the Danish mathematician and astronomer Olaus Rømer studied the times of the eclipses of Io, the innermost and thus fastest revolving of those four moons of Jupiter, and he found that the rhythm of its eclipses was imperfect. The discrepancies could amount to 22 minutes. But the irregularities of rhythm turned out not to be haphazard. As the earth in its orbital motion approached Jupiter, the eclipses came earlier and earlier than expected, and as the earth receded from Jupiter they came later and later. Rømer realized that this could be understood if light took some 22 minutes to

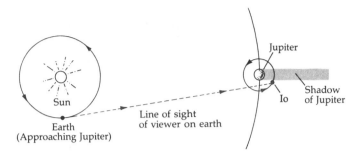

travel the length of a diameter of the earth's orbit. (According to modern values, the speed of light is close to 300,000 kilometers (186,000 miles)/second, and the diameter of the earth's orbit is some 300,000,000 kilometers (186,000,000 miles). It is a trivial matter to see that light takes about 1,000 seconds to travel the length of the diameter. Since 1,000 seconds is approximately 17 minutes, many books say that Rømer found a discrepancy of 17 minutes. But actually he was off by some 5 minutes.) With the diameter of the earth's orbit as estimated at the time, Rømer's idea implied that light must have a speed of some 210,000 kilometers (130,000 miles)/second. Few people were ready to believe that light, if it were not instantaneous and hence motionless, could move so extraordinarily fast—but Newton was one of the believers. Rømer presented his idea in 1675. More than 50 years were to pass before it was corroborated, and the corroboration came quite unexpectedly.

Box 4.1

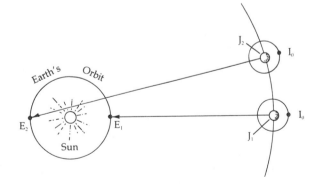

Here is Rømer's argument linking the irregularities of the observed eclipse times of Io, the innermost moon of Jupiter, to the speed of light. The eclipses were seen earlier than expected when the earth was nearer to Jupiter and later when the earth was farther from Jupiter. In the diagram E_1, J_1 are positions of the earth and Jupiter when they are closest to each other, and E_2, J_2 are their positions when they are farthest apart. When a moon of Jupiter emerges from eclipse, it suddenly becomes bright. But an observer on the earth does not see a brightening until the light from the suddenly bright moon reaches the earth. Light from J_1 to E_1 would have to travel the distance J_1E_1 while that from J_2 to E_2 would have to travel the distance J_2E_2, which differs from J_1E_1 by the diameter of the earth's orbit. Rømer attributed the lateness of the eclipses observed at position E_2 compared with those at position E_1 to the time needed for the light to traverse the extra distance equal to the diameter of the earth's orbit.

The ancients had pictured the stars as being like jewels fixed in a vast crystal sphere that revolved around the earth once a day. That is indeed how they appear to the naked eye. But, starting in the middle of the seventeenth century with the Frenchman Jean Picard, astronomers armed with telescopes noticed a strange yearly motion of the stars as if they were not firmly fixed in their settings on the celestial sphere. Early in the eighteenth century, the English astronomer James Bradley spent years studying the motion, at first with his friend Samuel Moleyneux and later alone. Let me describe the motion with the help of later knowledge.

Not one of the stars seems firmly set in the celestial sphere. All seem to move in tiny ellipse-shaped loops, in unison, each taking a year to travel its loop. The ellipses range from fat to thin but are all neatly lined up parallel to the plane of the earth's orbit, and their greatest diameters are all the same—some 40 seconds of arc, which is an amount that could be covered by the width of a not-too-fine hair held at arm's length.

Here was a heavens-wide, hair's-breadth ballet of the stars danced to a rhythm set by the earth. Bradley knew it had to be an appearance rather than a reality; the earth, no longer the center of the universe, could not command the heavens. One possibility was that the motion was due to what is called parallax, the perspective changes in the positions of stars as seen from a moving earth. For example, consider the apparent positions as seen from the earth of the star S in the diagram when the earth is at point A of its orbit and when the earth is at point B of its orbit. The lines AS and BS from the earth to the star tilt to the right and to the left, so when the earth is at A the star will seem to be displaced toward the right, and when the earth is at B the star will seem to be displaced toward the left. But the displacements found by Bradley and his friend were in an unexpected direction. Note that in the diagram the tilting lines AS and BS representing parallax displacements are on the page—at right angles to the instantaneous directions of the earth's motion, which are through the page. The Bradley displacements at each moment were in the same direction as that of the earth's motion, not at right angles to it. Thus they could not be due to parallax. (Parallax was not observed until 1838, more than a century later.)

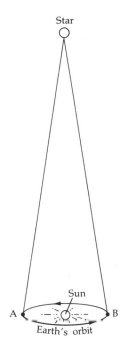

Star

Sun

A   B

Earth's orbit

The explanation of the strange stellar motions came to Bradley in September of 1728, and his letter to the Astronomer Royal, Halley, was read to the Royal Society of London the following January. The phenomenon, which is referred to as the aberration of light, deserves special attention not only because it confirmed that light is a journeying thing, but also for a further reason that will emerge later in this chapter. To get the feel of the idea, imagine yourself out with an umbrella in rain falling vertically on a windless day. If you stand still the rain will fall vertically relative to you, and to protect yourself you will have to hold your umbrella straight up. But if you start running, the rain will seem to be coming at you somewhat frontally and you will have to incline your umbrella accordingly, as indicated in the diagram. Now think of the rain

Standing   Running

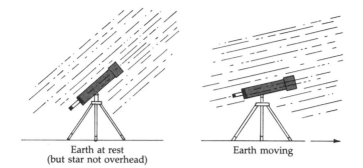

Earth at rest
(but star not overhead)

Earth moving

as analogous to light from a star and the umbrella as analogous to a telescope down which the star's light must pass if the star is to be observed. Because of the earth's motion, the telescope must point at each instant in a direction slightly inclined toward the direction of that motion. The star will therefore seem to be slightly displaced in that direction. And since the direction of our motion keeps changing as the earth moves in its orbit, the star will seem to be correspondingly displaced; it will appear to move over a year's time in a tiny closed loop that turns out to be an ellipse.

Box 4.2

Here, in outline, is Bradley's argument linking aberration with the speed of light. Following the custom at the time of the discovery of aberration, think of light as a stream of particles and, for the sake of simplicity, consider light coming down to earth vertically from a star directly overhead. To see the

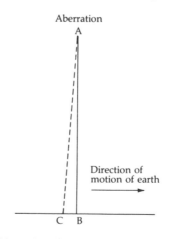

Aberration

general pattern of how aberration affects the observed position of stars, imagine the sun to be at rest, so that the earth's motion is just its orbital motion. We may do this because the effect of a reasonably steady solar

48

motion would be to shift the observed aberration pattern without obscuring the part of it arising from the earth's orbital motion. The earth's daily rotation can also be ignored, since it gives a much smaller effect than does the orbital motion. Draw a vertical line AB with the point B on the ground. If the earth were not moving, the vertically moving light that happened to pass through point A would strike the ground at point B. But with a moving earth the point B would have moved forward a little in the brief time it took the light to travel from A to the ground, and the light would therefore strike the ground at a point C a little behind B. The slope of the line CA, which is measured by the ratio of the length of AB to the length of CB, is equal to the ratio of the speed of light to the speed of the earth. Since, for the orbital motion of the earth, this ratio is approximately 10,000, we see that the length of the line CB in the diagram is highly exaggerated. Relative to the moving earth, the light would seem to be coming down not vertically but in a direction AC. The star would therefore seem to be not directly overhead as it actually was but a little displaced in the forward direction of the earth's motion. Since the earth moves in an orbit, its "forward" direction keeps changing, and therefore the star seems to trace out on the celestial sphere a miniature copy of that orbit. Essentially the same applies for all stars, except that the miniature copies of the earth's orbit become flattened by amounts that vary with different directions of the stars.

We see from this that the idea of journeying light could account in a general way for the observed effect. But how did it fit numerically? Since the amount of aberration depended on the ratio of the orbital speed of the earth to the speed of light, Bradley was able to determine how fast light travels. Rømer had found that light takes some 11 minutes to traverse a distance equal to the radius of the earth's orbit, but subsequent observations of the moons of Jupiter had considerably reduced this interval of time, some astronomers even setting it as low as 7 minutes. Bradley's aberration gave 8 minutes and 13 seconds, in striking agreement with the later Rømerian estimates. The corresponding speed comes to some 303,000 kilometers (188,000 miles)/second, which is close indeed to the modern value of 299,792 kilometers (186,272 miles)/second.

Much later, in 1849, the French physicist Armand Fizeau succeeded in measuring the speed of light not astronomically but terrestrially, there and back, reflected by a mirror between points 8 kilometers apart. He used a rapidly spinning toothed disk to time the light; the teeth blocked

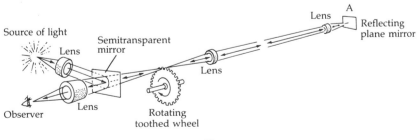

the passage of light and the gaps between them let light through. He adjusted the rate at which the disk spun until light that went through a gap in the disk could just make the trip of 8 kilometers and back in time to be blocked by, say, the very next tooth of the disk. Knowing the distance travelled by the light, the rate at which the disk spun, and the distance between the disk's teeth, he was able to compute the speed at which the light travelled. In 1862 Fizeau's friend, the French physicist Jean Foucault, was even able to measure the speed of light by an experiment wholly contained within a laboratory using rapidly spinning mirrors. And since then ever more subtle and accurate ways have been devised to determine light's speed.

What manner of thing was this light that moved so fast? Newton, arguing that it cast sharp shadows, believed that it consisted of particles. His contemporary, the Dutch physicist Christiaan Huygens, believed that light was a sort of wave. Newton's eminence was not the only reason that his particle theory of light came to be preferred to the wave theory. Genius that he was, he explained by his theory practically all of the properties of light that were known in his day, though in doing so he brought in wave concepts.

Starting in 1800, the English physician, physicist, and, later, Egyptologist, Thomas Young attacked the particle theory of light and proposed that light consists of waves. He gave compelling new arguments in favor of the wave theory. For example, light falling on light could produce darkness—one of a group of phenomena that Young referred to as the interference of light. There seemed no way to explain this phenomenon by means of the particle theory of light. One particle cannot cancel another. But interference posed no problem for the wave theory. Think of two sets of overlapping water waves. If the circumstances are such that at a given location the waves are always completely out of step—one at a crest when the other is at its deepest and vice versa in continued alternation—then at that place the two waves will cancel each other to produce an absence of effect; and with light this absence is darkness.

Box 4.3

When ocean waves strike a sea wall with two gaps, they create ripples that radiate from the gaps. At the gaps the ripples are always in step, cresting simultaneously. At a place like the one indicated in the diagram by a heavy dot, the ripples from the two gaps always arrive in step and thus combine to produce a greater wave amplitude. But at a place like the one indicated in the diagram by a small circle, the ripples arrive out of step, a crest of the ripple from the upper gap arriving when a trough from the lower gap arrives and vice versa; and when one ripple at that place is, say, three-quarters of the way up to its crest, the other is three-quarters of the way down to its trough. Thus at this place the two sets of ripples cancel and produce no disturbance. If we think of light waves passing through two slits, we see that they will give rise to patterns of light and darkness analo-

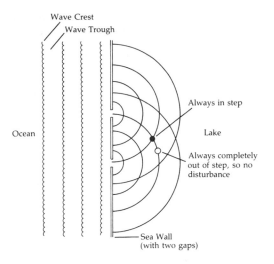

gous to the patterns produced by the ripples. Those patterns of light are referred to as interference fringes.

Young's idea was promptly ridiculed. But within a quarter of a century it had overwhelmed the particle theory. This transformation of outlook came about largely because of the dazzling researches of the French scientist Augustin Fresnel, starting in 1815.

Box 4.4

Physicists of the nineteenth century argued on the following grounds that light consists not of particles but of waves. A ray of light bends when

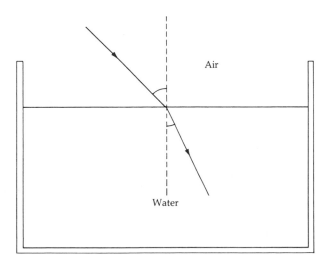

passing from air to water. Newton accounted in detail for this bending by postulating an attraction experienced by particles of light as they come extremely close to the water, which is more massive than the air. Since such a force would speed up the particles, it followed that the speed of light must be greater in water than in air.

The wave theory also accounted in detail for the bending, but by a different method. The waves, as they entered the water, were assumed to slow down and cause the change of direction by lagging. The speed of light in water would thus have to be slower than in air.

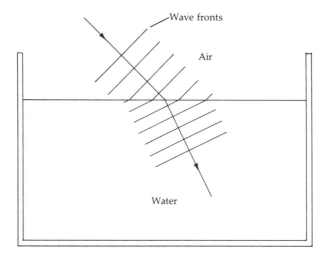

In 1850 Foucault succeeded in comparing the speeds of light in air and water experimentally. The speed was slower in water by the amount predicted by the wave theory.

The accomplishments of the wave theory of light were many and varied. Here is a relatively early sample: Consider a pinhole-sized source of light casting a circular shadow of a coin. According to the wave theory of light—but not the particle theory—the shadow of the coin would not be a dark circle but a dark circle with a bright spot at the center. According to the particle theory of light, a shadow arises because no light reaches the shaded area; according to the wave theory of light, a shadow arises because the light that reaches the shaded area from various points cancels itself out through interference. There is, however, an exception. The light does not cancel itself out at the exact center of a round shadow, and a spot of light thus arises. When this extraordinary prediction was verified experimentally, it naturally became a powerful argument in favor of the wave theory of light.

What is a wave? When a gust of wind strikes a wheatfield we see a wave speed across it. Each wheat stalk sways back and forth in the

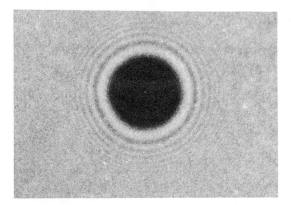

Because of the diffraction of light at a certain distance, the center of the shadow cast by an opaque disk shows a small bright spot (variously called Fresnel's, Arago's, or Poisson's spot).

direction of the wind, but no wheat stalk is uprooted; no wheat stalk travels from one end of the field to the other as the wave does. Again, if we have a rope straight before us and wiggle the near end left and right or up and down, we send a wave coursing along the rope even though the rope itself does not leave our hand to hasten after the wave. Waves do not transport, they transmit; and what they transmit are such things as energy and information.

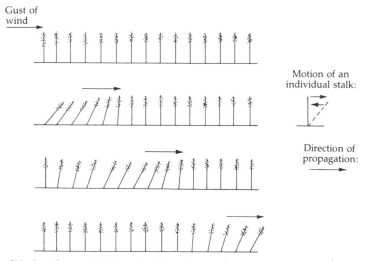

Side view of a gust of wind traveling through a wheat field, an example of *longitudinal* waves. Note that the stalks of wheat move in the direction of the propagating wave.

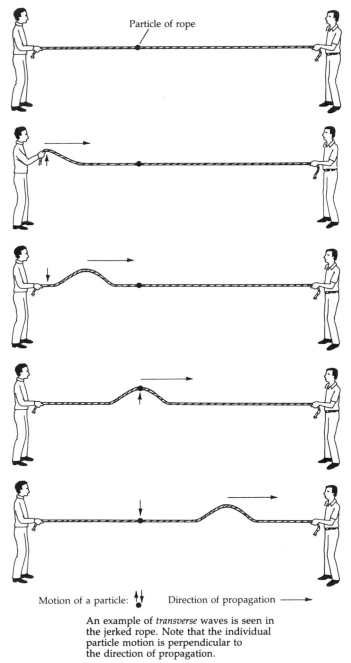

Particle of rope

Motion of a particle: ↑↓ •   Direction of propagation ⟶

An example of *transverse* waves is seen in
the jerked rope. Note that the individual
particle motion is perpendicular to
the direction of propagation.

54

If, as is the case with sound waves in air, the particles of the medium sway back and forth like the wheat tops in the direction of propagation of the wave, we say that the wave is longitudinal. If, however, like the motion of particles of the wiggled rope, the particles of the medium sway back and forth in a direction perpendicular to the direction of propagation of the wave, we say that the wave is transverse.

At first, Young and Fresnel—like Huygens in Newton's time—thought of light waves as analogous to sound waves, and thus as being longitudinal. But longitudinal waves could not account for the property of light that is called polarization. This property was known in Newton's time. Nowadays one can observe it by playing with strips of the polarizing material put out by the Polaroid Corporation or, indeed, with two

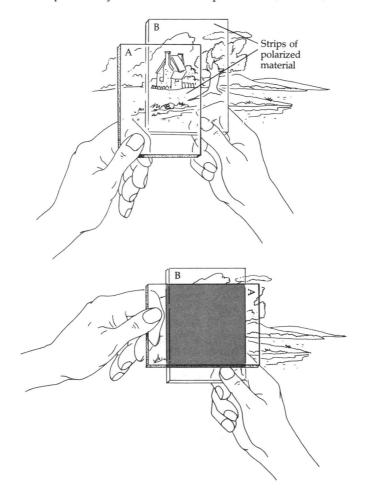

Strips of polarized material

pairs of polarizing sunglasses. Hold two of these strips (or lenses of the sunglasses) next to each other like adjacent cards in a deck, and light will pass through them freely; but if, keeping them flat against one another, you turn one through a right angle relative to the other, the pair will now block the passage of light. Newton accounted for the polarization of light by saying that light had "sides," which we may think of as meaning that the particles were not spherical but presented an elongated front as they came toward us. At first Young and Fresnel were at a loss to account for the polarization of light. Only after long puzzlement did it dawn on them that polarization could be neatly and simply understood in terms of transverse instead of longitudinal waves of light. While the idea of transverse waves of light proved highly successful, it had a drawback. Waves imply a medium in which to travel. No such medium was known for light, so scientists postulated one. This medium is called the luminiferous ether, or simply the ether. It must fill all space as far as the eye can see with the most powerful telescope, for if we can see an object, there must be an uninterrupted ether to carry the light waves from the object to our eyes. But there can be no transverse waves within a gas or a liquid. To provide the necessary lateral forces for a transverse wave within a medium, the medium would have to behave like an elastic solid. Moreover, an elastic-solid ether would have to have enormous rigidity per unit mass in order to be able to transmit waves with the prodigious speed of light. But Newton had accounted in detail for the observed motions of the planets. The cumulative effect of even a slight slowing down of their motion by the ether through which they were moving would quickly become evident. After all, astronomers had always found the motions of the planets to agree well with Newtonian predictions. But how could an all-pervading elastic-solid ether have no observable effect on the planetary motions?

There was no satisfactory solution, though many were offered that were extremely ingenious. Scientists learned to live with the problem. The wave theory of light was far too successful for them to give it up.

Box 4.5

The various phenomena associated with the polarization of light can all be understood in terms of transverse waves. For simplicity, take the case of polaroid strips and assume that the direction of their polarizing activity is parallel to the direction of their length. Polarized light is regarded as consisting of transverse waves all of which oscillate in the same direction. The three diagrams on the left show a transverse pulse of light polarized in the vertical direction, and the three diagrams on the right show a horizontally polarized pulse.

Vertically polarized light passes freely through the vertically polarized strip. Light polarized at an angle to the vertical has an amount of up-and-down motion, and this vertical component passes through the vertical strip and emerges as vertically polarized light. The amount of the vertical compo-

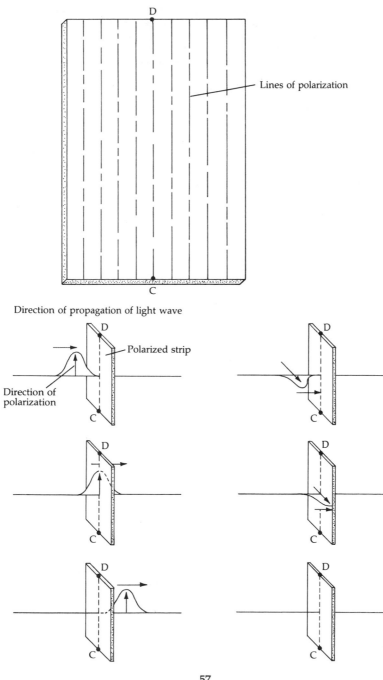

Lines of polarization

Direction of propagation of light wave

Polarized strip

Direction of polarization

nent decreases as the angle made with the vertical increases until the wave is horizontal, in which case there is no residue of vertical motion and the light is blocked by the vertical strip. This explains why the two crossed strips are opaque in their region of overlap.

Unpolarized light consists of a mixture of light waves polarized in various transverse directions. When unpolarized light passes through a horizontal strip it emerges horizontally polarized and is therefore blocked by a vertical strip.

Note how crucial is the transversality of the light waves in the above.

Bradley had explained the aberration of light using the particle theory. What if one used the wave theory? Then to account for the aberration of light, one would have to assume that the ether flows freely through matter—which was at least consistent with the need to have it offer no perceptible resistance to the motions of the planets. Suppose for the sake of argument that the ether near the earth was carried along by the earth. Then there would be no observed aberration. The reason is not hard to see. The amount of the aberration is found by subtracting the motion of the earth from the motion of the light. But the moving ether carried along by the earth would sweep the light waves along with it and add back the velocity of the earth, just cancelling the aberration. Thus because of the aberration of light, one must assume, to use Young's picturesque analogy, that the ether passes freely through matter much as the wind does through a grove of trees.

This free passage of ether through matter is extremely important. We recall that Newton had introduced absolute space in order to be able to speak of absolute rest and absolute motion, only to have his laws imply a principle of relativity that said that rest and uniform motion are relative. If the ether not only filled all space but, except for the ripples constituting light waves, was also unaffected by the bodies moving through it, then it could reasonably be regarded as being at rest in absolute space—indeed as constituting a sort of physical embodiment of Newton's absolute space. Therefore, by measuring how we are moving with respect to the ether, we could determine our absolute motion. Newton would have been pleased by this. From the new point of view, the disturbing Newtonian relativity would apply only to mechanical experiments. Optical experiments would yield the absoluteness that he had postulated right from the start.

Naturally, scientists began making experiments to measure the absolute motion of the earth—its motion relative to the ether. We have already remarked in connection with the motion of a bead on a straight but possibly moving wire that there are no stationary milestones or other markers cemented into space against which to recognize rest. Neither are there any such markers in the ether. But for our purposes the ether has an advantage over space: It transmits light waves, and, in

Standing   Running

Standing,   Running,
with wind   with wind

various ways, as we shall see, these waves can be expected to play the role of markers—moving markers, admittedly, but markers nonetheless—against which to measure absolute rest and absolute motion.

To see how this could be, consider a seemingly simple way to measure our velocity relative to the ether—a measurement that for a variety of reasons cannot be performed in practice. Let us consider the matter in terms of an analogy. Imagine a boat on a placid lake, and let a mist hide the shore because the ether has no shores. We wish to find how fast we are moving relative to the water, and for simplicity we shall assume that we are moving in the direction in which the boat is pointing. Our first idea is to look for floating buoys, but there are none visible. Our next idea is to throw our own buoys or lifebelts or bits of wood into the water and to note how we are moving relative to them. By this means we could solve the problem for the case of the boat on the water. But the problem of the earth in the ether cannot be similarly solved; we cannot throw "buoys" or other such markers into the ether and expect them to come quickly to rest—there is no friction in the ether. Therefore, let us forbid the use of floating objects thrown into the water from the boat. What

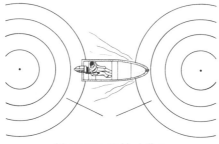

Waves generated by ballast

is there left for us to do? We take two lumps of ballast and hurl them into the water, one fore and one aft. Of course, they immediately sink. But when they hit the water they generate waves, one set coming at us from in front and the other from behind. We measure the speeds with which these waves pass us. Suppose the one from in front passes us at 11 kilometers/hour and the one from behind at 9 kilometers/hour. Then we conclude that the waves are actually travelling through the water at 10 kilometers/hour (the average of the two speeds) and that the boat is moving through the water at 1 kilometer/hour (half the difference of the speeds). From this we see how waves can indeed act as moving markers against which to measure our speed. In the case of our motion through the ether, the analogue of the two lumps of ballast would be two flashlights or other sources of light. If the speed of light is denoted by c and our speed through the ether by v, we would expect the waves from in front to pass us with speed $c + v$ and those from behind with speed $c - v$. The average is c, and half the difference is v. If we do not know the direction in which we are moving through the ether—which is surely the case—we perform the experiment in many directions and infer from the results the direction that will give the greatest value of v, which value will be our speed through the ether.

Naturally, scientists quickly became interested in the possibility of determining how the earth is moving. They already knew its orbital speed, but they did not know how fast the sun was moving and carrying with it its planetary flock. One experiment to measure the velocity of the earth was made as early as 1818 by the French scientist François Arago. It was what is called a "first order" experiment, meaning that its sensitivity was only enough to allow it to detect effects of the order of magnitude of v/c, where v is the speed of the laboratory and c the speed of light. If v is the orbital speed of the earth, v/c comes to 1/10,000.

The idea of the experiment can be grasped from the following considerations. The amount by which a glass prism bends a light ray passing through it depends, among other things, on the refractive index of the glass, and according to the wave theory of light the refractive index is the ratio of the speed of light in a vacuum to the speed of light in the glass. Suppose we have a glass prism at rest in the ether, and for simplicity let us take the speed of light outside it to be 3 units and inside it to be 2 units, so that the refractive index of the glass is $3/2$.

Now suppose that instead of the prism being at rest in the ether it is on the earth, and that the earth is moving with a speed of 1 unit toward the right. Then light coming from the right will, relative to the earth, have a speed of $3 + 1$ outside the prism and $2 + 1$ inside it. The refractive index for that light will then be changed from $3/2$ to $(3 + 1)/(2 + 1)$, that is, from $3/2$ to $4/3$. For light moving in other directions there will be different changes in the refractive index. From this we expect—and such an expectation also arises from a deeper, more mathematical investigation—that the amounts of bending of light rays by a prism on the earth will be affected by the motion of the earth.

When Arago made the experiment he found, to his intense surprise, that the earth's motion has no perceptible effect on the refractive index of glass. Arago told this unexpected result to his friend Fresnel, and Fresnel came up with an explanation that, for unexpected reasons, is one of the most intriguing in the history of physics. Fresnel assumed that the ether was present uniformly the same both inside and outside of matter, and that it flowed freely through matter. He then assumed that an additional amount of ether was permanently entrapped within the glass of the prism—and, indeed, within all transparent substances—the amount entrapped per unit volume depending on the refractive index of the substance. Thus there was more ether inside the prism than in a corresponding volume of free space. What happened when the prism was on the moving earth? Naturally, the entrapped ether moved bodily with the prism—one would hardly be calling it entrapped otherwise. But even as the entrapped ether was completely carried along by the prism, neither the prism nor its entrapped ether had the slightest effect on the omnipresent ether that was everywhere at rest.

For a wave of light entering the moving prism, the situation inside was complex. As seen from a position of absolute rest, there was the usual stationary ether and the additional entrapped, and thus moving, ether—with neither affecting the other at all. With this mixture of ethers, how fast would the light wave move? Fresnel said that it would move with its usual speed in stationary glass, supplemented by a speed equal to what one could call the average speed of all the particles of the two ethers present. It was a strange idea, but it did give a novel formula for the speed of light in moving glass and other media, and it accounted for the null result of Arago's experiment. Moreover, in 1851 Fizeau directly verified Fresnel's formula for the speed of light in moving media by measuring the speed of light in moving water. It was an experiment whose difficulty almost overwhelmed its extraordinary ingenuity.

Fresnel's formula applied to far more than the bending of light by moving prisms. He quickly realized that his formula had application to the aberration of starlight, for example. Recall that, mathematically, the aberration arises from subtracting the earth's velocity from the velocity of the incoming starlight. Suppose one filled a telescope with water. The light from the star would then move more slowly down the telescope. Subtracting the unchanged motion of the earth from the slowed motion of the light would yield a proportionally greater amount of aberration than before. But using his new formula, Fresnel was able to deduce that this first-order aberration experiment would show no effect of the water at all on the aberration. Much later, in 1871, this prediction was verified by the English astronomer George Airy.

Over the years, various other first-order optical experiments were performed to measure the velocity of the earth through the ether and thus the earth's absolute velocity. All gave null results. And it was ultimately shown mathematically on the basis of Fresnel's theory that all such experiments must indeed fail.

These failures to measure the earth's absolute speed, and thus also Fresnel's idea of entrapped ether, have about them the unmistakable scent of relativity. At first Fresnel's idea must have seemed strained and unlikely. But having observed its triumphs, we are now probably ready to welcome it. Let us, however, welcome it for the right reasons. The fact is that Fresnel's theory was utterly self-contradictory. Consider: Fresnel said that the amount of the entrapped ether depends on the refractive index. But the refractive index depends, among other things, on the color of the light. Thus if we were using red light, the amount of entrapped ether would have to be different from the amount needed if we were using blue light. But if the scheme is to make any sense, the amount of ether entrapped cannot change in this way. It has to be a single definite amount. If not, try to picture the etherial situation if we used white light, containing light of all colors.

We can see that Fresnel's idea is quite untenable. What shall we do, then? Laugh at it? Although that might be easy, it would reveal a misconception about the nature of science. Fresnel was facing problems that have no tenable solutions in a Newtonian setting, and in that setting inconsistency was almost inevitable. Let us admire Fresnel all the more for the intuition that directed his steps toward the solution of a problem that was only finally resolved with the advent of relativity. It took ability of the highest order for him to obtain his brilliant and farseeing results by such seemingly dubious means. Great science transcends logic.

Soon we will observe a Scottish physicist, James Clerk Maxwell, as he obtains even more important relativistic results by methods almost as unseemly. To understand as we watch him in action, we must first explore the subject of electromagnetism—a subject that will lead us directly to relativity.

Until the nineteenth century, there really was no such subject as electromagnetism. The ancients knew of magnetism through the attraction that the mineral called lodestone exerted on iron. They knew of electricity as a separate entity through the attraction exerted by amber on all sorts of substances after the amber had been rubbed. Not until the thirteenth century was there a beginning of significant progress in either subject.

More recently striking similarities were found between magnetism and electricity. Consider magnetism first. A magnet has two poles, north and south. Two magnetic poles of similar type—both north poles or both south poles—repel each other, whereas two magnetic poles of different types attract each other. The magnetic force between two magnetic poles, whether repulsive or attractive, is radial; that is, it acts along the line joining the poles. And the force varies inversely as the square of the distance between the poles.

Now consider electricity. It is believed to come in the form of tiny particles carrying electric charges, positive and negative. Two electric charges of similar type—both positive or both negative—repel each other, whereas two electric charges of different types attract each other. The electric force between two electric charges, whether repulsive or attractive, is radial; that is, it acts along the line joining the charges. And the force varies inversely as the square of the distance between the charges.

In view of such similarities, it is no wonder that scientists looked for a linkage between electricity and magnetism. But there seemed to be no connection. If an electric charge and a magnetic pole are placed at rest near each other, there is no observable effect of either on the other. Even so nature had given powerful hints of a relationship between electricity and magnetism; for example, lightning magnetized iron and affected compass needles. But the connection remained elusive until 1820. By that time scientists had realized that electric charges can flow quite freely through metals and other so-called conducting substances, even though most conducting substances are solids. A flow of electric charges is called an electric current.

In 1820, the Danish physicist Hans Christian Ørsted—a lifelong friend of the great storyteller Hans Christian Andersen—discovered that an electric current in a wire can deflect a compass needle.

The long-sought-after connection between electricity and magnetism had at last been found, and scientists could now understand why finding it had taken so long: The details of Ørsted's discovery were foreign to all their expectations. For example, the effect discovered by Ørsted was not a static one. Motion was crucially present—the electric current in the wire was a flow of electric charges, and if the flow ceased there would be no effect on the magnetic needle. As for the force between the current and the needle, it was in an utterly unexpected direction. The gravitational, electric, and magnetic forces between particles or poles at rest all acted along the line joining the particles or poles. But there was nothing the least bit radial about the influence of the electric current on the magnetic needle. In the diagram, let the central spot C represent a straight wire carrying an electric current downward into the paper. Then a magnetic needle at P will be deflected toward the right, as shown by the arrow, a needle at Q will be deflected toward the bottom of the page, a needle at R toward the left, and one at S toward the top of the page. The magnetic force due to the electric current is perpendicular to the current. To account for the effect of an electric current on a magnetic needle, Ørsted pictured the current as an axis about which raged a magnetic whirlwind.

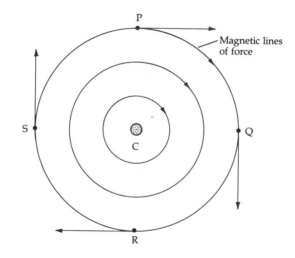

Ørsted's discovery immediately inspired the French physicist An-dré-Marie Ampère. In a classic theoretical and experimental investigation lasting three years, he discovered basic laws governing all known aspects of the new science of electrodynamics. So cogent and comprehensive was Ampère's work that Maxwell was to hail him as the Newton of electricity.

The phrase was apt in a deeper sense than Maxwell may have intended. For Ampère built his mathematical theory on what had come to be regarded as Newtonian concepts—action at a distance and purely radial forces—and Maxwell's own discoveries went counter to such concepts.

There was always the possibility that electric currents influenced magnetic needles through some previously unknown type of force. But Ampère showed that in every known respect electric currents gave rise to genuine magnetism. The reverse effect—magnetism giving rise to electric currents—is known as electromagnetic induction. It was discovered by the English experimenter Michael Faraday in 1831, and, about the same time, independently by the American experimenter Joseph Henry. The Russian experimenter H. F. E. Lenz was not far behind.

Faraday is of particular interest to us. As Einstein remarked, he is linked to Maxwell much as Galileo is to Newton.

Faraday, whose father was a blacksmith, ranks among the greatest experimental physicists of all time. He was born in 1791, and from age 13 until age 21 he worked as a bookbinder. In science he was largely self-taught, eagerly reading science books that came in to be bound. Fortunately, a perceptive customer took him to hear a popular series of science lectures given by the president of the Royal Institution in London. Faraday took careful notes and, at the urging of the customer, sent them, bound in leather, to the lecturer. The notes earned him his first job related to science. Starting as a lowly laboratory assistant at the Royal Institution in London, he was to become its director. His researches, powerful and voluminous, laid the foundation for our modern electromagnetic technology, yet one looks through them in vain for any mathematics other than occasional arithmetic, and even that is more apt to be expressed verbally than numerically.

Faraday's lack of facility in mathematics may seem unfortunate, but let us not jump to conclusions. Because of it he had to think about electromagnetic phenomena pictorially, and, as a result, his theories seemed at first to be naive and nonmathematical. Consider, for example, the simple static case of a horseshoe magnet attracting a tiny magnetic compass needle. For the mathematical physicists the essentials of the situation were the magnetic hardware and the inverse-square law of magnetic attraction between magnetic poles. For Faraday, however, the pieces of magnetic hardware were not of primary importance. Despite their immediate visibility, they were, for him, relatively inconsequential. He saw the horseshoe magnet, for example, as accompanied by an invisible presence that pervaded all space—a set of tentacles by which the magnetic poles pulled the needle and acted on other objects susceptible to magnetic forces. These tentacles could be rendered visible by means of iron filings. But they were there whether the iron filings were there or not. Faraday spoke of them as lines of force, and for him they were the prime magnetic reality. The region around the magnet was not empty. It was filled by these magnetic tentacles, always tugging, always crowding their neighbors and constituting what he called the magnetic field.

Similarly, he envisaged electric lines of force belonging to electric charges. For him they were the prime electric reality, and he spoke of them as constituting the electric field.

Are the lines of force in some sense there, or are they just an imaginative picture that helped the unmathematical Faraday to see a vague sort of order in his electromagnetic experiments? Compared with the

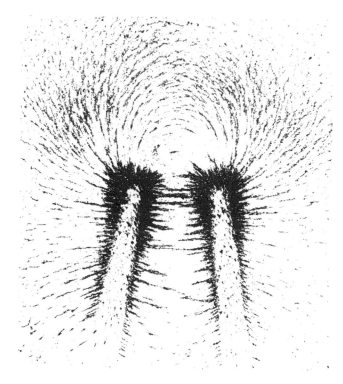

formulas of the mathematical physicists, the tentacles seemed naively imprecise. But they turned out to have a rich mathematical content that was spectacularly turned to account by Maxwell a quarter-century or so later—fortunately, while Faraday was still alive.

If we are not fussy about certain points that pertain to the calculus, it is easy to see in a simple case how the picture of tentaclelike lines of force could yield precise mathematical results.

Assume, with Faraday, that the pull exerted by a line of force is the same no matter what its length. Assume also—and here comes the calculus aspect—that the lines of force are so slender, so numerous, and so closely packed that we may think of them as smoothly present with no gaps between them even though they retain their own individuality.

Now consider the case of a single electric charge, with its lines of force—its tentacles—stretching out radially from it in all directions. Obviously all the lines of force will cross any imagined spherical surface having its center at the charge. Start with a sphere of unit radius. If a body having a small electric charge is placed on this sphere, there will be a certain number of the electric lines of force pulling on it, and the total pull on it of the central electric charge will be the sum of the individual pulls of the lines of force acting on it. Now double the radius of the sphere. Then its surface area will be quadrupled, and the lines of force as they cross it will be spread out more sparsely, with only a quarter as many as before available per unit area for pulling a charge on the surface. Thus the force's intensity will be only a quarter of what it was at the original distance from the central charge. If we triple the original distance the lines of force will be spread out to cover nine times the original surface area, so that the intensity of the force will be reduced to one-ninth of its original value. Thus, the seemingly unmathematical lines of force have led to the conclusion, known long before, that the electric force varies inversely as the square of the distance.

This, of course, was an especially simple case. But even in complicated cases involving many charges, the lines of force still embody the inverse-square law that lies almost unrecognizable amid the confusion.

That was not all. The lines of force proved effective in other situations as well. For example, in the case of electromagnetic induction, Faraday found his lines of force neatly apt for summing up his detailed researches in the form of a basic physical law. Essentially Faraday's law of electromagnetic induction said that to induce an electric current in a closed loop of wire, one must change the number of magnetic lines of force threading the loop of wire. It does not matter how one causes the number of magnetic lines of force threading the loop to change— whether by moving the magnet or increasing or decreasing its strength, or by moving the loop or distorting it, or by any combination of such means. As long as the number of magnetic lines of force threading the loop changes, there will be an induced electric current in the loop, and this induced current will be proportional to the rate of change of the numbers of lines threading the loop. Such was Faraday's discovery—a triumph for his seemingly unmathematical concept of lines of force as the very essence of electromagnetism.

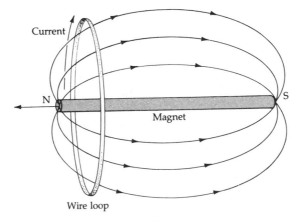

Current

N   S

Magnet

Wire loop

At a time when people were asking what kept bodies moving, Galileo told them to ask, rather, what brought bodies to rest or otherwise changed their motion. Faraday initiated a comparable revolution. At a time when people were concentrating their attention on the visible electromagnetic hardware, he told them to think, rather, of the rich, invisible content of the surrounding space—the electromagnetic field. And, as already mentioned, just as Galileo had a Newton to develop his ideas with dazzling mathematical skill and physical intuition, Faraday had a Maxwell to perform a comparable service.

Maxwell was born in Edinburgh in 1831, when Faraday was 40 years old. Thus Faraday was already an aging man when Maxwell began probing the mathematical content of his discoveries and ideas. The work required superb mathematical skill controlled by an audacious intuition, but Maxwell was more than equal to the task.

Maxwell was profoundly influenced by the Scottish physicist William Thomson, who was to become Lord Kelvin. At an early age Thomson had begun the job of giving mathematical form to Faraday's ideas by displaying a mathematical analogy between electric forces and the flow of heat in a solid.

Apparently at Thomson's suggestion, Maxwell postponed the mathematical study of electricity and magnetism until he had mastered Faraday's voluminous experimental researches on the subject. Thus he came to Faraday's work with a fresh mind and could see in it mathematical depths that lay hidden from other mathematicians.

Maxwell's first attack on the problem of electromagnetism was inspired by the work of Thomson. Maxwell probed the mathematical relationship between electric lines of force and fluid flow in a highly fictional fluid of infinite extent. Inside it he imagined fluid spouting from electric charges of one sign and being sucked back into oblivion by charges of the opposite sign. By noting the pressures built up, he could account precisely for the attractions and repulsions of the charges. A similar analogy held for magnetism. Finding no reasonable fluid mechanism that could account for electromagnetic induction, Maxwell simply decreed that his lines of flow behaved in such a way as to give rise to Faraday's rule about the lines of force threading the closed loop of wire. The range of electromagnetic phenomena that Maxwell was able to bring within the province of his new mathematical approach was extraordinary. Yet at this stage he was merely exploring a new mathematical analogy. On this score he was quite explicit, stating in the article in which he set forth his work, "I do not think it contains even the shadow of a true physical theory; in fact, its chief merit as a temporary instrument of research is that it does not, even in appearance, *account for* anything" (Maxwell's italics). Note the presence, both here and in the earlier work of Thomson, of a medium—the counterpart of Faraday's idea of a field. Faraday had seen the field—the lines of force teeming in space—as a prime reality. Any theory faithful to Faraday's vision would have to be a theory of an all-pervading field.

In 1861, some six years after the investigations described above, and having at last met Faraday, Maxwell returned to the problem of electromagnetism. This time he went beyond a purely mathematical analogy and boldly sought a mechanical model for an ether pertaining to and indeed shaping electromagnetic phenomena. Ørsted, we recall, had vaguely imagined an electric current to be surrounded by a sort of whirlwind that affected magnetic needles. Ampère, after thorough analysis, had concluded that magnetism was a secondary effect arising from circulating electric currents. It is interesting to contrast all this with Maxwell's new idea. He began with pure magnetism, which, following the views of Faraday and Thomson, he regarded as something rotatory. He pictured it, therefore, in terms of an ether consisting of small "molecular vortices," or spinning fluid globules, whose axes of rotation lay along the lines of magnetic force; if the directions of spin of the vortices were reversed the magnetic lines of force would reverse their directions. Thus in his basic assumption Maxwell differed from Ørsted, whose whirlwinds were large and raged around electric currents; and he differed from Ampère not merely in starting with magnetism but in shunning a disembodied, mathematical action at a distance in favor of the concept of a field.

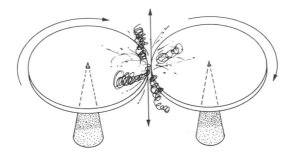

Having set up this basic vortical picture, Maxwell worried about friction. In a smooth magnetic field one would want the molecular vortices over a wide region to be rotating in the same direction. But in that case, the rims of neighboring vortices would be moving in opposite directions where they met. Maxwell needed neighboring vortices capable of spinning in the same direction without friction. Designers of machinery had already encountered that sort of need, and had met it by introducing small "idle wheels" between cogwheels. In the diagram, note how the idle wheel, by turning counterclockwise, lets both cogwheels turn clockwise. With this in mind, Maxwell separated neighboring molecular vortices from each other by tiny spherical particles, as shown in the diagram on the next page that he himself gave by way of illustration. Note the unsatisfactory six-sided vortices. The diagram does help to indicate how the spherical particles could maintain a single layer of particles between vortices, but does it not show adjacent spherical particles in friction-creating contact, thus acting counter to the reason for their inclusion?

What of the spherical particles? Were they merely a sort of lubricant to lessen friction and thus assuage a guilty scientific conscience? By no means. Having introduced the spherical particles under vortical duress, Maxwell assigned to them a central new role, saying that "their motion of translation constitutes an electric current," and that the idle wheels "play the role of electricity."

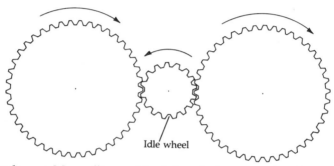

Idle wheel

Such was Maxwell's model of the electromagnetic ether: vortices attended by particles. Maxwell was candid about his "imaginary system of molecular vortices," saying in the paper in which he presented his theory:

> The conception of a particle having its motion connected with that of a vortex by perfect rolling contact may appear somewhat awkward. I do not bring it forward as a mode of connexion existing in nature, or even as that which I would willingly assent to as an electrical hypothesis. It is, however, a mode of connexion which is mechanically conceivable, and easily investigated, and it seems to bring out the actual mechanical connexions between the known electro-magnetic phenomena; so that I venture to say that any one who understands the provisional and temporary character of this hypothesis, will find himself rather helped than hindered by it in his search after the true interpretation of the phenomena.

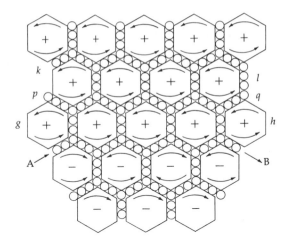

For all that Maxwell's model seemed hopelessly bizarre, there was magic in it. From it he derived a set of equations for the electromagnetic field that accounted in rich mathematical detail not only for Ørsted's discovery and all its elaborations at the hands of the illustrious Ampère, but also for Faraday's law of electromagnetic induction and, indeed, for all the main features of electromagnetic phenomena then known. Once having found his field equations, Maxwell made them the basis of his theory. The vortices and idle wheels could now be retired. They had served his purpose—and served it well.

One crucial, and controversial, step taken by Maxwell in obtaining his equations was the introduction of what he called the displacement current. Consider a piece of a substance, say glass, that does not conduct electricity. Because it is a nonconductor, we might expect that there could not be an electric current in it. But Maxwell, influenced by Faraday's views, said otherwise. The molecules of the glass can be regarded as holding electric charges captive within them. If an electric force is applied to the glass, these captive charges will strain at the leash and become slightly displaced. The brief motions that result in the displacements, being a momentary flow of electric charges, count as an electric current. It is called a displacement current. As Maxwell neatly put it, "a displacement current is not a current but the commencement of a current." But by varying the electric force one could vary the displacement and, thereby, generate a varying displacement current of long duration.

Leading scientists among Maxwell's contemporaries, including his mentor and friend Thomson, had great difficulty accepting Maxwell's bold ideas. The displacement current was particularly troubling. In view of the key role of the glass above, one might have difficulty accepting Maxwell's assertion that there would be displacement currents in empty ether. And yet, since Maxwell endowed his electromagnetic ether with various physical properties one could hardly object to his declaring it capable of sustaining displacement currents, even though one might have misgivings. A greater cause for concern among Maxwell's contemporaries seems to have been the role of the displacement current as an actual current. For example, since total current—regular plus displacement—always flowed in closed loops, scientists wondered how Maxwell's theory could account for the existence of local accumulations of electric charges.

Box 4.6

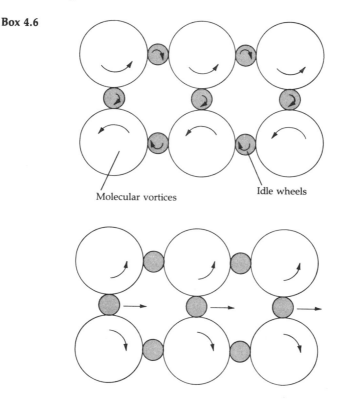

Molecular vortices

Idle wheels

It is possible to see without mathematics how Maxwell's molecular vortices account for the magnetic effect of an electric current. The upper diagram shows vortices and spherical particles in linked rotations pertaining to a uniform magnet field. Start with no motion at all, so that initially there is neither magnetism nor electric current. Let us now try to create an electric current flowing from left to right in the two-dimensional diagram. To do this we move the inner row of three particles bodily to the right. This sets the molecular vortices spinning, and we now have motion, as in the second diagram, which shows that magnetism is present. It shows more. Note that the upper vortices are turning counterclockwise, and the lower ones clockwise. This means that above the current the magnetic lines of force point toward us and below the current they point away from us. If we had considered a three-dimensional diagram, we would have seen that the magnetic lines of force circle the current, which is what Ørsted had observed more than forty years before and had vaguely envisaged as a magnetic whirlwind raging around the electric current.

Justifiably or not, the displacement current was felt to be inconsistent with other Maxwellian concepts of electric charge and current. The general unease about Maxwell's ideas can be sensed from the following remarks of the brilliant German experimental and theoretical physicist Heinrich Hertz, especially if one bears in mind that Hertz was a great admirer of Maxwell. Here is what Hertz wrote: "Unfortunately the word 'electricity' in Maxwell's work obviously has a double meaning. . . . If we read Maxwell's explanations and always interpret the meaning of the word 'electricity' in a suitable way, nearly all of the contradictions that at first are so surprising can be made to disappear. Nevertheless, I must admit that I have not succeeded in doing this completely, or to my entire satisfaction."

Let us recall that Maxwell did not believe in his molecular-vortex model of the electromagnetic ether and stopped using it once it had led him to his electromagnetic field equation. Something analogous occurred with the image that lay behind the concept of the displacement current. With the advent of relativity, scientists realized that without the mathematical term representing the displacement current, Maxwell's equations would be in conflict with Einstein's theory. When the mathematical term was included, Maxwell's equations were in full conformity with that theory. The image behind the displacement current, for all its problems, had done its work and could be retired. Maxwell, of course, could not know of such developments that lay in the future. He had achieved something more extraordinary than he could have imagined. He had found electromagnetic equations that fit superbly into the relativistic structure of space and time that superseded the Newtonian framework in which Maxwell, like Fresnel before him, had had to work. Little wonder, then, that in the Newtonian setting of absolute space and absolute time Maxwell's equations caused puzzlement.

Because of the displacement current, Maxwell's equations could be written in a form in which the symbols pertaining to electricity and magnetism entered in almost identical ways, creating a fascinating formal symmetry. Intimately tied to this symmetry between electricity and magnetism was a dazzling mathematical conclusion. Although Maxwell seems only to have glimpsed the symmetry, he was able to find the conclusion: that there should be electromagnetic waves, and that these waves would have to be transverse. As for their speed of propagation, the equations said that it must be equal to the ratio of two different units

73

of electric charge; one unit had to do with the electric force between two electric charges at rest, the other unit with the magnetic force between two electric currents—electric charges in motion. This ratio had already been determined by means of an electromagnetic experiment. Within the limits of experimental error, it turned out to be equal to the speed of light. But the experiment essentially involved neither light nor waves. As Maxwell pointedly remarked, "The only use made of light in the experiment was to see the instruments." The fact that the experiment gave the speed of light had been widely regarded as a coincidence. But Maxwell, having found it to be associated with his theoretical transverse electromagnetic waves, declared light to be electromagnetic.

Maxwell's theory showed electricity and magnetism to be so tightly and symmetrically interrelated as to constitute different facets of a single entity. Moreover, it linked this unified electromagnetism with light, so that light was no longer something separate but merely a particular manifestation of electromagnetism. The electromagnetic ether was identified with the luminiferous ether—the bearer of light waves.

Maxwell's theory was a superb unification of optics and electromagnetism. Yet for several years it remained half-believed and half-doubted. Maxwell died in 1879 at the age of 49, too soon to taste the joy of vindication.

Not until some nine years after Maxwell's death was his theory convincingly confirmed by direct experiment—the experimenter being none other than Heinrich Hertz, who was quoted above. Hertz generated invisible electromagnetic waves and showed that they behaved as Maxwell had predicted. As a result, Maxwell's theory came into its own in scientific circles. But conceptual difficulties still remained. Hertz, for example, who had said enthusiastically, "From the outset Maxwell's theory excelled all others in elegance and in the abundance of the rela-

The Electromagnetic Spectrum

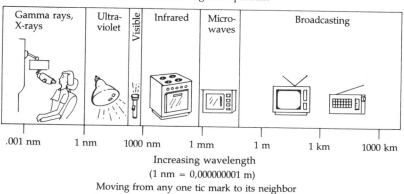

Increasing wavelength
(1 nm = 0,000000001 m)
Moving from any one tic mark to its neighbor
to the right increases the frequency by a factor of 1000.

tions among the various phenomena that it included," nevertheless remained puzzled by what he called "Maxwell's peculiar assumptions or methods." Peculiar they were indeed, but they were great science. Moreover, they led to applications in everyday life. For example, radio and television signals are transmitted by means of electromagnetic waves. One wonders, though, how many people who use radios and television sets are aware of the work of Maxwell—or even of his name.

Quite apart from his work on the electromagnetic field equations, Maxwell had pondered the question of the earth's motion through the ether and had suggested that the velocity of the earth could be found by timing the eclipses of Jupiter's moons. Rømer had used these eclipses in his work on the speed of light. Maxwell proposed a different use for them. Suppose for simplicity that relative to the ether the solar system is moving toward the right. Then when the earth and Jupiter are lined up like this

Earth   Jupiter

the light signals bringing news of the eclipses will be traveling toward an approaching earth and will therefore arrive earlier than expected. However, when the earth and Jupiter are lined up like this

Jupiter   Earth

the eclipse signals will be traveling toward a receding earth and will therefore arrive later than expected. By observing the variations of the rhythm of the eclipses as seen on the earth, one should be able to infer the velocity of the solar system, and thus also of the earth, through the ether.

In 1879, the American astronomer D. P. Todd, having completed careful tables of those eclipses, sent a set to Maxwell, who sent back a detailed letter of thanks. A few months later, Maxwell died. Realizing that Maxwell's letter now took on a special sentimental and historical significance, Todd sent it to the Royal Society of London. It was published in their *Proceedings* and republished in the British scientific journal *Nature*, which then, as now, had a worldwide circulation among scientists. As a result, Maxwell's letter came to the attention of the American physicist Albert Michelson.

In the letter, Maxwell remarked that theoretically there is an effect of the earth's motion on the there-and-back speed of light as measured in the laboratory, but in practice one could not use it to measure the earth's velocity, because, as he said, it "is quite too small to be observed." Since to detect the orbital speed of the earth by this method one would need to be able to measure a time interval of about a thousandth of a millionth of a millionth of a second, Maxwell's view is understandable. But he had reckoned without the ingenuity and experimental audacity of a Michelson. Michelson declared the effect "easily measurable" and promptly set about planning to measure it.

Michelson, who was born in Poland in 1852, emigrated to the United States as a young child, his family probably taking the step in order to avoid anti-Semitism. He became an ensign and later an instructor at the U.S. Naval Academy, where he became an expert physicist and an authority on the experimental investigation of light. In 1880 he took a leave of absence to study in Berlin, where he developed an instrument of exquisite sensitivity, now called an interferometer, in order to carry out the experiment mentioned by Maxwell and thereby measure the earth's absolute motion.

The design of the interferometer exploited the facts that the speed of light is enormous and that the wavelength of visible light is very small—around $1/20,000$ of a centimeter. As already mentioned, the experiment proposed by Maxwell required the measurement of a time interval of about a thousandth of a millionth of a millionth of a second. So incredibly small a time interval could not be measured directly, so Michelson measured the distance travelled by light in that time instead. That distance turns out to be roughly the size of a wavelength of visible light. Michelson's interferometer used the light and dark interference fringes produced by a pair of light beams, which in effect allowed him to measure that small distance by using the wavelength of the light as his yardstick.

The sensitivity of the interferometer had its drawbacks. When Michelson set up his apparatus in the laboratory of the Physical Institute in Berlin, the vibrations caused by traffic in the street played havoc with his readings. So he moved to the smaller town of Potsdam and set up his apparatus underground in the quiet room housing the telescope mount at the observatory. Even so, when someone stamped on the pavement 100 meters away, the delicate apparatus was thrown temporarily out of adjustment. And, as Michelson ruefully remarked, this happened with an interferometer purposely made insensitive.

Box 4.7

The idea of Michaelson's experiment has justly received the acclaim of physicists. The aim of the experiment was to detect the ether wind as follows. Let OA and OB be of equal length and at right angles to each other. If there is no ether wind, the time taken for light to travel from O to a mirror at

A and back will be equal to that taken for light to travel from O to a mirror B and back.

Now suppose that the earth is carrying the apparatus through the ether toward the right—and in order to see what is happening, suppose the speed of the earth to be a sizable fraction of the speed of light. To hit the moving mirror B, one must aim the light not perpendicularly to OA but a little forward. (This is closely akin to aberration.) So, relative to the ether, the paths of light will be as in the diagram, and as a result the two journeys will no longer take equal times. The purpose of the Michelson-Morley experi-

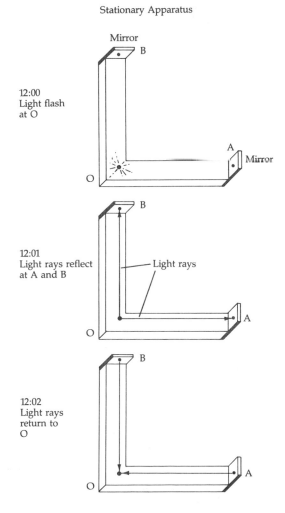

Stationary Apparatus

Mirror

B

12:00
Light flash
at O

A
Mirror

O

B

12:01
Light rays reflect
at A and B

— Light rays

O

A

B

12:02
Light rays
return to
O

O

A

Both light rays reach O at the same time.

Moving Apparatus

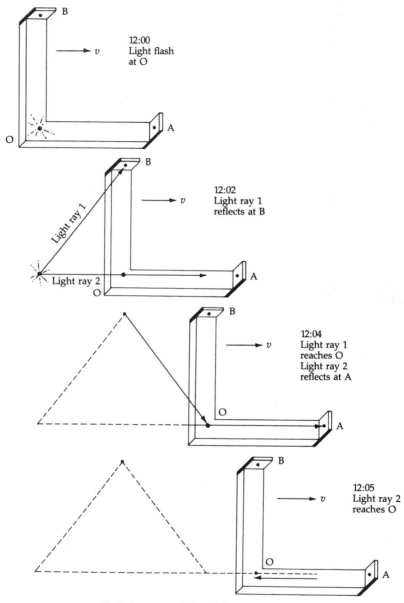

12:00
Light flash
at O

12:02
Light ray 1
reflects at B

12:04
Light ray 1
reaches O
Light ray 2
reflects at A

12:05
Light ray 2
reaches O

The light rays reach O at different times.
The drawing is done for the case $v = \frac{3}{5}c$.

ment was to measure the difference of the times and thus to determine the velocity of the earth through the ether.

Stripped to its bare essentials, the apparatus consists of a half-silvered mirror at O and regular mirrors at A and B. In the experiment a beam of light from a source S is split by the half-silvered mirror at O, one part going through to mirror A and back and the other part being reflected to B and back. If the light journeys are of exactly the same length, the light waves will return in step, producing an interference pattern consisting of a bright spot at the center surrounded by alternating dark and light fringes when the light is viewed through the telescope at T.

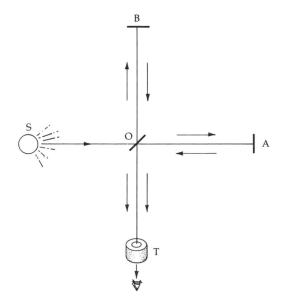

If one light journey becomes slightly larger than the other, the interference pattern will be shifted laterally. This can be easily seen in the special case in which one journey is longer than the other by half a wavelength. For in that case at the former bright places where the two beams were in step they would now be completely out of step, one beam at crest when the other was at trough *vice versa*, thus cancelling to produce darkness. Similarly the former dark places would now be bright, the interference pattern thus being shifted in the present special case by half the distance between neighboring bright fringes.

Now suppose that the motion of the earth affects the lengths of the light paths, and thus the times taken, as predicted. Then turning the apparatus would cause the interference pattern to move horizontally, and by noting the orientations that produce the greatest displacement of the interference pattern, one could find the path along which the earth moves through the ether as well as the earth's speed along that path (though not, it turns out, which way, forward or backward, the earth is moving along the path). However—and this was the great surprise—the experiment gave no sign of an ether wind—no indication at all that the earth was moving. Yet the experiment was capable of measuring a time interval something like that taken by light to travel a distance of a hundredth of a wavelength.

Michelson published a confident account of his 1881 experiment, claiming that there was no indication of any motion of the earth. But he had made a theoretical error in his calculations that had caused him to overestimate the size of the expected effect of the earth's orbital motion on the interferometer readings. The correct value was only half of what he had thought. Consequently the experiment turned out to be inconclusive. Nevertheless it did show that with only a small increase in sensitivity, the effect of the earth's absolute motion, if it existed, should be measurable.

In 1887, at the Case School of Applied Science in Cleveland, Ohio, in collaboration with the chemist Edward Morley, Michelson repeated the experiment with major refinements that greatly increased its sensitivity and accuracy, and this time the lack of effect of the earth's motion on the interferometer readings had to be accepted.

But acceptance of the null result of the experiment led to a problem. The aberration of light had shown that the ether must flow freely through matter. Thus there must be an unshieldable ether wind relative to us as we and the earth speed through the ether. Various first-order experiments had failed to detect this ether wind, but their failure could be accounted for by means of Fresnel's idea of entrapped ether alongside free ether. The Michelson-Morley experiment was a second-order experiment, that is, it looked for an effect having the size not of v/c but of $(v/c)^2$, which was much smaller, and it therefore lay beyond the reach of Fresnel's formula.

Thus according to the theory of aberration there should be an ether wind, but according to the Michelson-Morley experiment there was none.

The Special Theory of Relativity

CHAPTER 5

IN 1902, SOME FIFTEEN YEARS AFTER THE
Michelson-Morley experiment, Michelson wrote:

> The experiment is to me historically
> interesting, because it was for the solution
> of this problem [of motion through the
> ether] that the interferometer was devised.
> I think it will be admitted that the
> problem, by leading to the invention of the
> interferometer, more than compensated for
> the fact that the experiment gave a
> negative result.

We can understand Michelson's disappointment. He had hoped to be the first to detect the earth's motion through the ether, but all he had accomplished, as he saw it, was to establish that the ether, instead of flowing freely through the earth, was carried along with it. This view had been put forward by the English physicist George Stokes in 1845, long before the Michelson-Morley experiment. He thought that the ether was completely carried along near the surface of the earth but not higher up. To account for the aberration of light, he had had to assume that the ether was an incompressible fluid in which there were no vortices. But Lorentz, among others, had pointed out that such an ether could not match the motion of the earth all over its surface. It thus could not resolve the problem of the undetected ether wind.

Soon after the Michelson-Morley experiment was performed, the Irish physicist George FitzGerald proposed in his lectures that its null result could be accounted for if the motion of an object through the ether caused the object to contract in the direction of its motion. Specifically, the object would have to become shorter by a factor $\sqrt{(1 - v^2/c^2)}$, where v is the speed of its motion through the ether and c is the speed of light. For everyday speeds this would be a negligible shrinkage. Even the earth moving rapidly around the sun would be contracted by a mere 6 centimeters or so—the length of a small blade of grass. For speeds close to the speed of light, however, the contraction would be considerable, and at the speed of light all lengths in the direction of the motion would contract to zero. The proposed contraction would exactly cancel the second-order effect that had inspired the Michelson-Morley experiment. But most of FitzGerald's scientific friends who heard of his idea of a contraction laughed at it, to his distress.

However, in 1892 the Dutch physicist Hendrik Anton Lorentz proposed the idea of a contraction independently—and published it. There is a charming incident connected with this event. Two years after publishing the contraction idea, Lorentz heard of FitzGerald's having proposed it. Wishing to give proper credit, Lorentz asked FitzGerald if he had ever published the idea. FitzGerald wrote back saying he had not, thus conceding priority of publication to Lorentz. On his part, Lorentz hastened to give public credit to FitzGerald, not only stating that FitzGerald had thought of the contraction independently but implying that he might well have thought of it first. It was a gentlemanly act.

But the story has an unexpected final twist. As the American scientist Stephen Brush discovered in 1967, FitzGerald had been mistaken! In 1889 he had outlined his idea in a letter to the American journal *Science.* But *Science,* because of financial difficulties, temporarily discontinued publication, and FitzGerald, believing it defunct, assumed that his letter had remained unpublished. He apparently did not know that *Science* started up again and that his letter had appeared in it later in 1889, three years before Lorentz published the contraction idea, and only two years after the Michelson-Morley experiment.

Thus FitzGerald had the priority. But it was Lorentz who vigorously pursued the matter. He was the world's greatest expert on Maxwell's electromagnetic theory. By 1895 he had produced a powerful extension and simplification of the theory. In it the ether, except for such things as the ripples caused by electromagnetic waves, was completely stationary, the electromagnetic field equations being regarded as valid for laboratories at rest in the unmoving ether. The question arose of how the equations would change if one went over to a laboratory moving uniformly relative to the ether. Some preliminary remarks will be helpful in describing the steps by which this key question came to be answered.

To specify the location of a point on a page, we can draw graph-paper lines on the page, select an origin, O, and through it draw an x axis and a y axis as shown. We can then assign coordinates (x, y) to the point. For example, in the diagram the point P has the coordinates (3, 1). The coordinates of the origin are (0, 0).

In three dimensions we use a third coordinate, z, which may be taken to be the height of the point above the page. For example, if

4 units above the page, there is a point directly over P, it has $x = 3, y = 1$, and $z = 4$, so that its coordinates (x, y, z) are $(3, 1, 4)$. The three-dimensional coordinates of the origin are $(0, 0, 0)$.

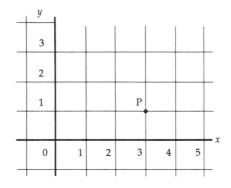

In going from two to three dimensions, we replace the flat graph-paper grid by a three-dimensional scaffolding with one intersection point, O, selected as origin and three mutually perpendicular beams through O selected as x axis, y axis, and z axis. The scaffolding is conceptual rather than material. It can thus be thought of as extending throughout space. When such scaffoldings are equipped with clocks for measuring time, they will be called frames of reference, or reference frames.

Since we shall be comparing measurements made by two observers whose frames of reference are in uniform motion relative to each other, it is fortunate that the extensive scaffoldings are conceptual—otherwise they would wreck one another because of their relative motion. To distinguish the coordinates x, y, z of one of the observers from the x, y, z of the other, it is customary to write the latter with primes, as x', y', z'. The following relations, together called a Galilean transformation, tell us how to transform from the coordinates used by one of the observers to those used by the other:

$$x' = x - vt, y' = y, z' = z,$$

or equivalently

$$x = x' + vt, y = y', z = z',$$

where v is the speed of the primed observer relative to the unprimed one. A hypothetical example will indicate the role and importance of transformation equations. Suppose that the unprimed observer experimenting in his frame of reference found a law of motion stating that every free particle always moves so that $x = y$—an absurd "law" but instructive nonetheless. How would this "law" appear to the other observer? To find out, we apply the Galilean transformation, that is, we use the transformation equations to replace the unprimed coordinates in the "law" by primed coordinates. Since $x = x' + vt$ and $y = y'$, the equation $x = y$ is transformed by the Galilean transformation into the equation

$$x' + vt = y'.$$

Box 5.1

In Newton's absolute space and absolute time, a Galilean transformation links the coordinates belonging to two frames of reference that are in uniform motion relative to each other. The frame with unprimed coordinates is often regarded as being at rest, with the other frame moving uniformly relative to it with speed v. For simplicity, it is customary to consider a special case in which the scales of distance and time are respectively the same in both frames, the primed coordinate axes coincide with their unprimed counterparts at time $t = 0$, and the relative motion is in the common x and x' direction.

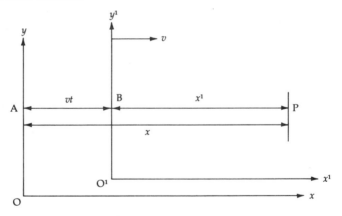

The z and z' axes have been omitted in the diagram, and for clarity the x and x' axes have been slightly separated. For any point P the x coordinate is the distance AP and the x' coordinate the distance BP. At time t, because the two frames are separating at speed v, the distance AB is equal to vt. But BP = AP − AB. Therefore $x' = x - vt$, and this relation, coupled with $y' = y$, $z' = z$, is called a Galilean transformation. It converts unprimed into primed coordinates. With an eye to later developments, one may append the relation $t' = t$, which says that the readings of clocks in the primed reference frame agree with the readings of clocks in the unprimed frame—all of them measuring absolute time. Thus the Galilean transformation may be written

$$x' = x - vt, \; y' = y, \; z' = z, \; t' = t.$$

These relations are mathematically equivalent to the relations

$$x = x' + vt', y = y', z = z', t = t',$$

which convert primed into unprimed coordinates.

Apart from the primes, this differs from the previous, unprimed equation by the presence of the term vt, which contains the quantity v and thus depends on the relative speed of the two reference frames. Since v does not enter the unprimed equation, the unprimed reference frame is somewhat special, and we may think of it as being at rest in absolute space or in the ether. The primed reference frame is then moving uniformly in absolute space or the ether. Suppose that the primed observer in the laboratory with the primed frame of reference made experiments to measure the coordinates x' and y' of a free particle at some particular time t. Suppose the observer found $x' = 4$, and $y' = 10$ when $t = 1$. Then by inserting those numerical values in the transformed equation—the primed equation—the primed observer would obtain $4 + v = 10$, showing that $v = 6$. Thus by making measurements solely of primed quantities, the primed observer will have measured the absolute speed of the primed laboratory. So if the "law" $x = y$ had been valid, speed would have been absolute.

Instead of the spurious "law" that we invented for illustrative purposes let us now consider Newton's laws of motion. They may be expressed in terms of x, y, z, and t in the unprimed frame of reference, which we are taking to be at rest in absolute space. When we apply a Galilean transformation in order to find the equations—and thus the physical laws—that hold in the uniformly moving primed reference frame, it turns out that, except for the primes, the primed equations are exactly the same as the unprimed equations. Since v does not enter the primed equations, no amount of experimental measurement of primed quantities will yield the value of v; one cannot find the value of v from an equation in which v does not appear. In view of our experience in the airplane in steady flight, this absence of any effect of v in the uniformly moving primed frame should not surprise us. It is a mathematical way of expressing the principle of relativity that Newton deduced from his laws.

While Newton's equations belong to the science of mechanics, Maxwell's equations express the laws of electromagnetism. Let them be accepted as valid in a frame of reference at rest relative to the ether. When a Galilean transformation is applied to them the primed equations turn out to contain the quantity v in the first-order combination v/c and in the smaller second-order combination v^2/c^2. Because of the presence of v in the primed equations, an observer in a laboratory fixed in a moving primed frame of reference ought to be able to find the value of v by making electromagnetic (which includes optical) measurements; that is, he ought to be able to detect the ether wind arising from his motion through the ether. But, as we know, first-order experiments that attempted to detect this ether wind had failed to do so.

At this stage Lorentz made a major discovery. First he made a mathematical alteration in the Galilean transformation, replacing the relation $t' = t$ by the more complicated relation $t' = t - vx/c^2$. Because through the presence of the coordinate x, this new quantity t' depended on the location, he called it local time, to distinguish it from true universal time t. He then showed that if one applied this altered Galilean transformation to the unprimed Maxwell equations and made certain adjustments, the primed equations had the same form as the unprimed equations except for terms containing v only in the small second-order combination v^2/c^2. The absence of larger, first-order terms meant that, to the first order, any electromagnetic experiment in a laboratory moving uniformly through the ether when interpreted in terms of local time would give the same result as a corresponding experiment in a stationary laboratory interpreted in terms of true time. This showed that a first-order electromagnetic experiment could not distinguish between stationary and uniformly moving laboratories. Lorentz thus deduced that first-order electromagnetic experiments to detect whether a laboratory was moving relative to the ether would fail to detect any motion. In particular, he obtained Fresnel's formula for the speed of light in moving media without the self-contradiction of conflicting amounts of entrapped ether that had marred Fresnel's derivation of the formula.

The French mathematician, expositor of science, theoretical physicist, and philosopher of science Henri Poincaré took a broader view, and many of his utterances were strikingly prophetic. As early as 1895 he had objected to the patchwork approach to the problem—Fresnel's idea for explaining away the null results of first-order experiments to detect an ether wind, and a contraction of lengths for explaining away the second-order Michelson-Morley experiment. What, asked Poincaré, if other experiments to detect an ether wind gave null results? Would we invent separate makeshift explanations for each in turn? He felt that a single explanation should suffice for all. In 1904 he even spoke of "a principle of relativity" and suggested that there should be a new mechanics in which nothing could exceed the speed of light.

By 1904 second-order experiments different from the Michelson-Morley experiment were indeed also showing no sign of the earth's absolute motion. Poincaré's strictures against a patchwork approach greatly influenced Lorentz, who in 1904 published a key paper entitled "Electromagnetic Phenomena in a System Moving with Any Velocity Less than That of Light." The restriction on the velocity entered basically because at the speed of light lengths would contract to zero.

In this paper Lorentz did avoid a patchwork approach to the explanation of all of these results, as Poincaré had desired. Lorentz incorporated the FitzGerald-Lorentz contraction of lengths into his transformation formulas and made a corresponding change in his definition of the local time. This gave him what is now called the Lorentz transformation—a name bestowed on it by Poincaré in 1905.

Box 5.2

In the theory of relativity the Galilean transformation is replaced by what is called the Lorentz transformation. The Galilean transformation is $x' = x - vt$, $y' = y$, $z' = z$, $t' = t$. With "local time" it becomes

$$x' = x - vt, \, y' = y, \, z' = z,$$
$$t' = t - vx/c^2.$$

The Lorentz transformation differs from this only by having a multiplying factor (denoted by β) in two of the equations:

$$x' = \beta(x - vt), \, y' = y, \, z' = z$$
$$t' = \beta(t - vx/c^2),$$

where $\beta = 1/\sqrt{(1 - v^2/c^2)}$.

Note that the denominator of β is the factor by which lengths are shortened in the FitzGerald-Lorentz contraction.

The Lorentz transformation was so named by Poincaré in 1905, though the mathematical transformation had already been used by the English physicist Joseph Larmor in 1898, and something closely akin to it by the German physicist W. Voigt as early as 1887.

As for the crucial importance of the Lorentz transformation, recall first that if one applied the Galilean transformation to the unprimed Maxwell equations, the resulting primed equations had additional first-order terms involving v/c and second-order terms involving v^2/c^2. Recall too, that if one applied the altered Galilean transformation using local time and made certain adjustments, the first-order terms disappeared but second-order terms were still present. If one applies the Lorentz transformation and makes certain adjustments, the primed equations, except for the primes, turn out to be exactly the same as the unprimed equations. No additional terms of any sort—first order, second order, or any other—enter the transformed equations. Since the unprimed frame of reference was taken to be at rest in the ether with the primed one moving uniformly relative to it, one could deduce that a wide variety of electromagnetic experiments designed to detect motion relative to the ether, including the Michelson-Morley experiment, would give null results provided one's motion was unaccelerated. (Lorentz had made an error in applying the transformation, but his work sufficed to explain the Michelson-Morley experiment and most other experiments one might have devised to measure motion through the ether. His error was corrected by Poincaré, who put Lorentz's work in its final form in 1905.)

To see how the argument went with regard to the impossibility of detecting motion through the ether by means of electromagnetic experiments, consider the case of the Michelson-Morley experiment. Since that experiment was designed to detect motion relative to the ether, it would have to give a null result in a laboratory stationary in the ether. Thus in such a laboratory the two beams in their round-trip journeys would arrive back at their starting point at the same true time. Therefore, according to Lorentz's work, in the same experiment performed now in a laboratory moving with constant velocity through the ether, the same must be true in terms of local time when account is taken of the FitzGerald-Lorentz contraction. But if the two beams arrive at the starting point at the same local time, they will also arrive there at the same true time, since *at a given location* in the moving laboratory equal values of local time correspond to equal true times. Similarly, the true times of initial departure of the beams will be the same. Thus the two beams will take equal true times for their journeys even in a uniformly moving laboratory, which means that there will be no displacement of the interference pattern as the apparatus is rotated and thus no sign of any motion of the earth through the ether. An analogous argument would apply to any experiment in which local times were compared at fixed locations.

This brings us back to the airplane with which we started, and the fact that from his laws of mechanics Newton deduced what we have been calling the Newtonian principle of relativity: that inside a vehicle moving uniformly relative to absolute space, there is no mechanical effect of the motion.

Optical and other electromagnetic effects of the motion initially were expected to be detectably present in the interior of a vehicle moving uniformly relative to the ether which is why electromagnetic experiments were performed to detect the earth's motion relative to an ether at rest in Newton's absolute space. Because of the null results of those experiments, and the work of Lorentz in accounting for those results, we might well be tempted to extend Newtonian relativity to include electromagnetism. But the situation is not as simple as it may seem.

In the Newtonian case, in transforming from the unprimed reference frame at rest to the primed reference frame in uniform motion, the absence of interior mechanical effects of the motion—that is, the absence of v in the primed Newtonian equations—was a result of using a Galilean transformation. In contrast, the analogous absence of interior electromagnetic effects—that is the absence of v in the primed Maxwell equations—was a result of using a Lorentz transformation with its built-in contraction of lengths and its strange mathematical substitute for true time.

And now, at last, Albert Einstein enters the story. He was born in Ulm, Germany, on March 14, 1879—the year in which Maxwell died. His early years were not what one might expect of a man who was to rank among the greatest scientists of all time. He did not learn to speak until he was three years old. He hated the German schools, with their strict discipline and their emphasis on rote learning. A teacher once told him he would never amount to anything. He became a school drop-out. He failed the entrance examination at the Zurich Polytechnic Institute. When he did enter, he cut his classes and alienated his professors. He and his fellow student Marcell Grossmann became firm friends, and if Grossmann had not loaned his lecture notes to Einstein, it is quite possible that Einstein would not have been able to graduate. He found cramming for his final examinations so distasteful that not until a year afterward could he bring himself to think about science again. Upon graduation he could find no academic post; after occasional tutoring and teaching jobs, in 1902 he was fortunate to obtain, with Grossmann's help, a position in the Swiss patent office in Bern as Engineer: Third Class.

Of course, the picture was not as one-sided as that. When he was 5 years old he was given a magnetic compass by his father, and the sight of the totally enclosed needle striving toward the north with no visible cause filled him with a sense of awe and wonder that remained with him for the rest of his life. At age 12 he felt a comparable ecstasy on coming across a textbook on geometry, and it is possible that the sight of so many striking theorems being developed from simple axioms helped lead him later to base his scientific theories on simple and general scientific principles that play a role similar to that of the axioms in geometry.

Young Einstein quickly became fascinated by science. He viewed the universe not only with a burning curiosity but also with the sort of awe, wonder, and admiration that one is more apt to associate with the religious mystic than with the scientist. Much later in life he said that when judging a scientific theory, his own or that of someone else, he would ask himself whether, if he were God, he would have made the universe in that way. A theory lacking the cosmic beauty one would expect of God was not acceptable to Einstein, except as a stopgap for want of anything better. We shall be seeing cosmic beauty in his theories of relativity.

Einstein persistently cut classes at the Polytechnic Institute in order to study the masterworks of science on his own and to perform experiments in the laboratory. Thus in a significant sense he was self-taught.

At the patent office Einstein kept his passion for science burning brightly. And in the fabulous year 1905 his genius burst into spectacular flower. For a comparable flowering of genius one would have to go back to the plague years in England when Newton meditated in the quiet of Woolsthorpe.

The first of Einstein's 1905 papers was one of his most audacious. Like all his scientific papers of that year, it was published in the distinguished German research journal *Annalen der Physik*. When Einstein was awarded the 1921 Nobel Prize in Physics, the only item of his work specifically mentioned in the citation was a formula derived in this first paper of 1905. But the paper was not about relativity.

Some five years before Einstein's paper, the German physicist Max Planck had been faced with puzzling new experimental data about the glowing of certain hot bodies. To account for these new data, he proposed that matter does not absorb or emit energy smoothly, as had always been thought, but in discrete amounts that he called *quanta*. To appreciate how revolutionary this was, consider the analogy of a children's swing. Since the distance through which it swings depends on its energy, Planck was proposing something analogous to saying that a swing can go 1 meter, 2 meters, 3 meters, and so forth, but not anything in between. The new theory accounted admirably for the experimental data. But not even Planck dared to take the quantum hypothesis seriously at the time. Indeed, he spent the next dozen years unsuccessfully trying to find a way to avoid it. The only person to take it seriously was the patent examiner, Einstein. In the first of his 1905 papers he proposed something related that was even more drastic. Though fully aware of the overwhelming evidence in favor of the wave theory of light, he gave cogent reasons for concluding that light should nevertheless be regarded as somehow consisting of particles. Some twenty years were to pass before this drastic proposal became generally accepted by scientists.

The second of Einstein's 1905 papers presented a new way of determining the dimensions of molecules. The third dealt with specks of dust and the like in a liquid. If the liquid consisted of molecules in agitated motion, as pictured in the molecular theory of heat, their bombardment of the specks should cause the specks to execute an incessant, irregular, microscopic zigzag motion. The Scottish botanist Robert Brown had noticed such a motion in 1827. When Einstein's formula for this Brownian motion was verified experimentally by the French scientist Jean Perrin in 1908, important scientists who had not believed in atoms were at last persuaded that atoms do indeed exist.

The fourth of the 1905 papers, submitted a mere thirteen weeks after the first, was entitled "On the Electrodynamics of Moving Bodies," and it set forth what we now call the special theory of relativity. Almost simultaneously Poincaré submitted for publication a long paper that contained much of the mathematical detail of Einstein's paper, and also important mathematical developments going beyond it. And yet the credit for the creation of the theory of relativity must go to Einstein. For example, since Lorentz and Poincaré based their arguments on the detailed structure of electromagnetic theory, their results had the status of an outgrowth of that theory. Einstein, however, deduced the Lorentz transformation from two general principles. He showed thereby that not the Galilean but the Lorentz transformation expresses universal relations having to do with the behavior of time and space themselves; and that the Lorentz transformation thus necessarily applies not just to electromagnetism but to all of physics. Although Einstein used the same mathematical transformation equations as Lorentz and Poincaré, his use of them was based on radically new concepts of the behavior of space and time.

Einstein began his paper by discussing the electromagnetic induction of a current in a loop of wire by means of a magnet. He stressed that the amount of current depends on the relative motion of the loop and magnet, and not on their absolute motion through the ether. But, continued Einstein, Maxwell's theory gives quite different physical explanations according as the loop is at rest in the ether with the magnet moving or the magnet at rest with the loop moving. If the magnet moves, it creates an electric field that gives rise to a current; if the loop moves and the magnet does not, there is no longer an electric field. Einstein spoke of the energy of the electric field in the first case, emphasizing its presence as a physical entity that cannot be ignored. He said that these different explanations of the current that arises when a magnet and loop of wire are in relative motion and "the unsuccessful attempts to discover any motion of the earth relative to the [ether]" suggest that there is no such thing as absolute rest. It is striking that Einstein here did not mention the Michelson-Morley experiment specifically, an omission that has led to debate as to whether he was acquainted with it at the time.

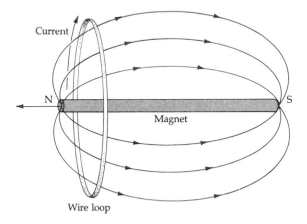

Einstein next proposed two principles as the foundation of his theory. The first, which he called the principle of relativity, said in effect that if we are in an unaccelerated vehicle, its motion has no effect on the way things happen inside it. This reminds us of Newtonian relativity, but Einstein was making it a basic principle valid not just for mechanical phenomena but also for electrodynamical and optical phenomena, and indeed for phenomena of all types. Another way of stating the principle of relativity is that the laws of physics are the same in all unaccelerated reference frames.

The second of the two principles in Einstein's paper said that the motion of light is not affected by the motion of the source of the light. Nothing, it would seem, could be more orthodox and obvious. For if a source of light generates light waves in the ether, once the waves are launched they are no longer linked to their source; they are on their own, moving at the rate set by the elastic properties of the ether.

There are various remarks to be made about this second principle. For instance, if it is so obvious, how could it turn out to be part of a revolution—especially when the first principle is also a natural one? Moreover, if light consists of particles, as Einstein had suggested in his paper submitted just thirteen weeks before this one, the second principle seems absurd: A stone thrown from a speeding train can do far more damage than one thrown from a train at rest; the speed of a particle is not independent of the motion of the object emitting it. And if we take light to consist of particles and assume that these particles obey Newton's laws, they will conform to Newtonian relativity and thus automatically account for the null result of the Michelson-Morley experiment without recourse to contracting lengths, local time, or Lorentz transformations. Yet, as we have seen, Einstein resisted the temptation to account for the null result in terms of particles of light and simple, familiar Newtonian ideas, and introduced as his second principle something that was more or less obvious when thought of in terms of waves in an ether. If it was so obvious, though, why did he need to state it as a principle? Because, having taken from the idea of light waves in the ether the one aspect that he needed, he declared early in his paper, to quote his own words, that "the introduction of a 'luminiferous ether' will prove to be superfluous."

We see in all this the workings of an extraordinary intuition. The beautiful thing about Einstein's cunningly chosen pair of principles is that each by itself seems harmless, yet the two together form an explosive mixture destined to rock the very foundations of science. Moreover, because of their simplicity it is possible to make key deductions from them with minimal use of mathematics, although Einstein usually chose a more mathematical route in his published papers.

Here is a sample of the revolutionary consequences that follow if we accept both principles. Imagine that you and I are in unaccelerated spaceships in the depths of space somewhere near a star. For simplicity, assume that my spaceship is at a fixed distance from the star and that yours is speeding toward the star at one-fifth the speed of light, as indicated in the diagram, which is drawn as if the page were at rest

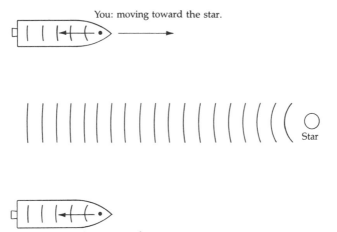

You: moving toward the star.

Star

Me: at rest with respect to the star.

relative to my spaceship and the star, and thus not relative to your spaceship.

In what follows, you and I will make similar experiments inside our respective spaceships. We shall also look outside our spaceships. The principle of relativity applies to the *interior* experiments and requires that similar interior experiments in your spaceship and mine will give similar results.

I light a lamp in the forward part of my spaceship and measure the speed of the light waves as they move toward the rear. You perform the same sort of experiment in your spaceship. By Einstein's second principle, which says that the motion of light is independent of the motion of its source, the light waves from the star, the light waves from the lamp in my spaceship, and the light waves from the lamp in your spaceship will all keep abreast of one another as they move toward the rear. So when I measure the speed of the waves from my lamp, I am also measuring the speed with which the light waves from the star go by me. And the same holds for you. Since you are rushing toward the star at one-fifth the speed of light, we are likely to conclude that you will find the light from the star, and thus also the light from your lamp, rushing toward you not at the speed of light as measured by an observer stationary with respect to the star—me—but with that speed augmented by your own speed of approach to the star. The light from the star would then pass you with a speed one-and-one-fifth times the speed with which it passes me. Therefore the interior light from your lamp would also pass you with a speed one-and-one-fifth times the speed with which the interior light from my lamp passes me. But this conclusion has to be false. It conflicts with Einstein's first postulate, the principle of relativity. For in measuring the speed of the light from our respective lamps, you and I are performing identical *interior* experiments and should therefore get identical results. If I find the speed of waves from my lamp to be c kilometers per second, so too must you. And, because of the exterior aspect of the experiment, you must therefore find the light waves from the star passing you at this same speed c—despite the fact that you are rushing toward the star.

We can look at the situation from a slightly different point of view. If you were moving toward the star with a speed one-fifth that of light, and found the light waves from the star to be passing you with the speed of light augmented by your own speed—one plus one-fifth times the speed of light—you would also have to find a speed of one and one-fifth times that of light for the waves from your lamp. But in that case, by sending light from a lamp in the opposite direction in your spaceship, you would find the speed of the waves to be the speed of light diminished by your speed—one minus one-fifth, or four-fifths, of the speed of light—and half the difference between the forward and backward speeds would be your velocity, which the principle of relativity forbids you to find. Therefore, even though you are travelling toward the star with speed $c/5$, you will have to find the light from the star and the light from the lamps passing you with speed c.

It is certainly a shock to discover that no matter how fast one rushes toward a source of light or away from it, the light waves from it will nevertheless go by with the same speed c. But there are many more shocks to come—so many that it is hard to know which to tell of next.

Suppose I bet that I can speed up so as to move relative to you with a speed as great as the speed of light. Then I shall lose my bet. Here is the reason. Imagine a race between light and me, allowing me as good a flying start as I wish in order to gather speed before the prow of my spaceship comes abreast of yours. The instant the prows are abreast, you send out light waves and the race is on. We have already seen that no matter how fast I move, the light waves will always move relative to me with the same speed. Therefore, since I shall never be able to catch up with the light waves, you will see me trailing behind them. So, no matter how hard I try, I shall not be able to speed up so as to move relative to you—or to any other unaccelerated observer—with a speed as great as that of light. The result holds true not just for me but for any material object whatsoever that has intrinsic mass. In relativity, the speed of light is the speed limit.

As we have seen, Einstein's two principles have surprising consequences concerning the motion of light. But their consequences concerning the idea of simultaneity are even more surprising—and those conse-

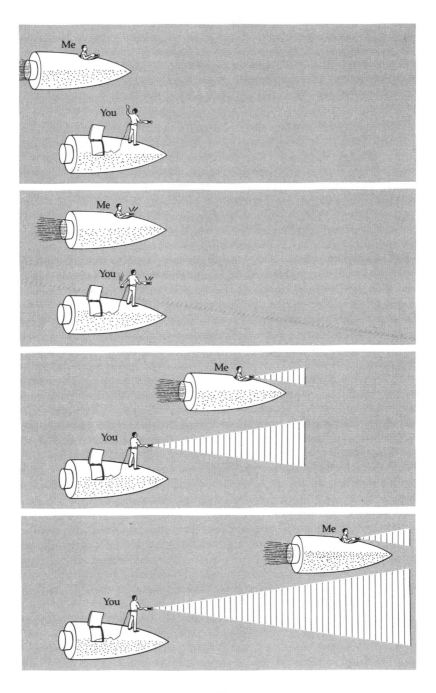

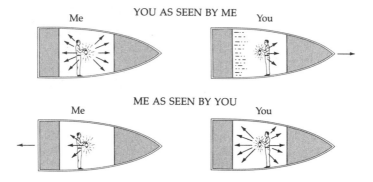

quences are crucial to Einstein's theory. They are part of the revolution he brought about in our concepts of time and space.

Einstein assumed that if two events happened at the same place, there would be no problem as to whether they were simultaneous or not. But he questioned the accepted idea of the simultaneity of events happening at different places. Flash a lamp exactly in the middle of the interior of your spaceship, so as to send light waves fore and aft. Then, since the light waves travel with the same speed in both directions, you will declare that the light waves will reach the front and the back of your spaceship simultaneously. If I similarly send light waves from the center of my spaceship, I will make a similar claim. But what will I think of your claim? Since your spaceship is moving forward relative to mine, I will see the back of your spaceship moving toward your rearward light waves and the front of your spaceship moving away from your forward light waves. So I will observe the rearward light waves in your experiment reaching the back of your ship before the forward waves reach its front. Thus the fore and aft arrivals of your waves, which you said occur simultaneously, do not occur simultaneously according to my observations.

This disagreement is extraordinary in itself. But even more extraordinary is the fact that the disagreement is a reciprocal one. With respect to your spaceship, mine is moving backward, so that according to your observations the forward light waves in my experiment reach the front of my ship before the rearward ones reach the back, and therefore the arrivals of my waves are not simultaneous according to you even though they are simultaneous according to me.

The principle of relativity requires that we here treat each other as equals. We may not decide to believe "you" rather than "me" or "me" rather than "you." We have to conclude, therefore, that if two observers are in uniform relative motion, then spatially separated events that are simultaneous for one are, in general, not simultaneous for the other, and vice versa. Simultaneity is relative.

Note how the relativity of simultaneity plays havoc with Newtonian ideas. In terms of absolute time we could decide at once whether two spacially separated events were simultaneous or not. They were simultaneous if they occurred at the same absolute time. In relativity, with simultaneity relative, we can say that time itself is relative.

In formulating his theory of relativity, Einstein had to be particularly clear about how to specify the location and time of a momentary event at a point—a point event. As was customary, he used a conceptual three-dimensional scaffolding for obtaining the spatial coordinates (x, y, z). At each intersection of the scaffolding he envisaged an accurate clock, all these myriads of clocks being synchronized with a master clock at rest in the scaffolding. Then the time, t, of a point event would be the time shown by the particular clock that was situated at the location of the event. So far, all is so simple as to seem overcautious. But we have not yet told how the clocks were to be synchronized, and it was there that the power of Einstein's intuition shone forth. Here is the essence of the method that he proposed. It will seem almost obvious—and that is part of the beauty of it.

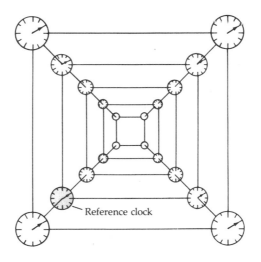

Reference clock

Imagine you and me with our scaffolding and clocks in our usual spaceships. I have a clock C at a fixed point on my x axis, and I wish to synchronize it with the master clock at my origin O. I send a flash of light from O to C, where it is reflected back to O. I then adjust the hands of C so that the time taken for the light to travel from O to C comes out to be the same as the time taken for the light to travel the return journey from C to O. As thus adjusted, clock C is declared to be synchronized with the master clock at O.

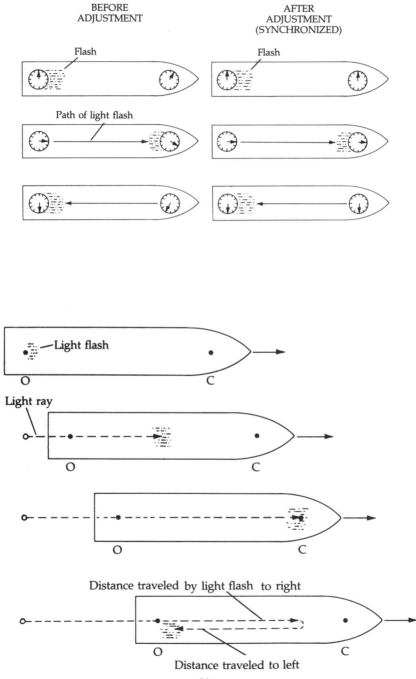

For example, suppose that the clocks give readings in billionths of a second and that the light left O when the master clock read noon, reached C when *that* clock read 4 billionths of a second after noon, and returned to O when the master clock there read 6 billionths of a second after noon. According to the clocks, the outward journey of the light took 4 billionths of a second and the return journey only 2. I conclude that clock C is not synchronized with the clock at O. The adjustment needed is simple. I turn back the "hands" of clock C by 1 billionth of a second. According to the clocks, the outward and return journeys of the light will now each take 3 billionths of a second. I therefore declare clock C synchronized with my master clock at O.

You, in your spaceship, perform a similar synchronization. And there still seems to be nothing of moment in all this. But observe what happens when I take note of your synchronizing activities and you take note of mine. Because you are moving relative to me, I see your light signals going unequal distances there and back. So the very fact that your clocks assign equal times for the two journeys shows that, relative to me, your clocks are definitely not synchronized.

But, relative to you, I am the one who is moving, and since you see my light signals travelling unequal distances, we find that, relative to you, my clocks are the ones that are not synchronized.

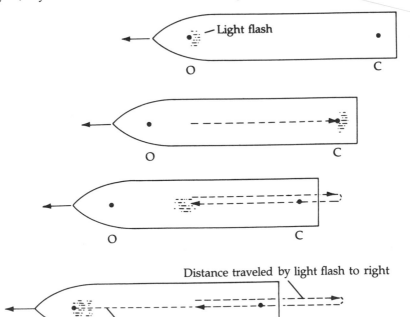

Distance traveled by light flash to right

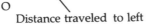

Distance traveled to left

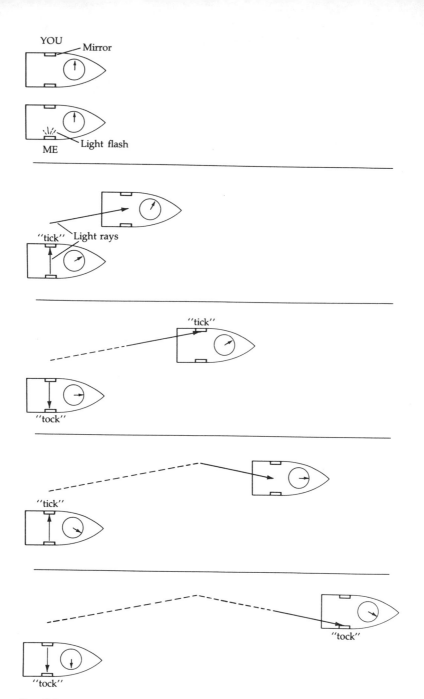

Your light clock as seen by me. Note that for the figure my clock runs twice as fast as yours. The speed of light is the same for both clocks. Consequently, since equal intervals separate successive frames, all dark arrows have the same length.

We are both right. According to the theory of relativity, there is no universal "now" as there was with Newton's absolute time. Suppose I synchronize all my clocks with my master clock and you do the same with yours. Then we will each have a means of assigning four coordinates, x, y, z, t and x', y', z', t' respectively to any point event, the t and t' being the readings of our respective clocks at the location of the event. If according to my synchronized clocks two point events at different places occur at the same time, I will legitimately say that the events are simultaneous, although, in general, you will legitimately disagree. Correspondingly, when you declare two spatially separated events simultaneous, I will be the one to disagree. We have thus arrived once more, but by a different route, at the conclusion that simultaneity is relative.

The relativity of simultaneity does more than play havoc with a Newtonian doctrine; it plays havoc with our everyday notions of the way time behaves, and because of that, since time is so basic, it gives rise to seemingly incredible developments, some but by no means all of which we have already encountered.

For example, suppose that we mount two mirrors so that they are parallel and face each other. Then if we let light be reflected to and fro between them, the light will be "ticking" off equal intervals of time, and therefore the light and the two mirrors will constitute a clock. It is not a clock designed for ordinary everyday use. But by its very simplicity it will help us explore the behavior of time according to the special theory of relativity. We may refer to it as a light clock.

We each place one of these light clocks in our spaceship and look at the behavior of the light clocks as seen by both of us. Consider the situation as seen by me. Because you are moving relative to me, the "ticking" light rays of your light clock will not be vertical for me. They will slope and strike the moving mirrors. According to my observations, the distances traveled from tick to tock to tick in your light clock are greater than those in mine. Since light travels with the same speed c relative to both of us, I will find that the time intervals ticked off by your light clock are longer than those ticked off by mine. Thus although the two light clocks are of identical construction, I will maintain that your light clock goes at a slower rate than mine. As it turns out, I will see your clock slowed by a factor of $\sqrt{(1 - v^2/c^2)}$, where v is our relative velocity and c is, as always, the speed of light. Suppose, for example, that we were moving relative to each other at four-fifths the speed of light. Then since $\sqrt{(1 - 4^2/5^2)} = \frac{3}{5}$, I would find that your clock would tick only three times while mine ticked five times.

Box 5.3

To derive the formula for the relativistic slowing of clocks, let us, as usual, have coordinates with x axes along the same line and with our relative motion in the common x direction with constant speed v. Also, as usual, let us both have coordinates using the same distance and time scales.

In deriving the time-dilatation formula, we shall make use of light clocks oriented perpendicular to our relative motion. The first step, therefore, is to make sure that our respective measurements of distance agree in directions perpendicular to our relative motion. That is important because a change in length could alter the ticking rate of a light clock. Essentially, we wish to show that, for each point, the transformation linking my coordinate y to your coordinate y' is just $y' = y$. The method of proof is to show that the assumption that y' is not equal to y leads to a contradiction.

If I find by measurement that you are moving relative to me with speed v toward the right, then you will find by measurement that I am moving relative to you with the same speed v toward the left. That is so because in these measurements we are doing essentially the same thing under the same circumstances—each measurement is the mirror image of the other. By symmetry, any effect on lengths in a direction perpendicular to the direction of the relative motion cannot depend on whether the motion is to the right or to the left, provided v is not changed. Using the symbol LT to stand for "is less than", let us suppose, to obtain a contradiction, that for a particular value of v your distance measurement to a given point is less than mine, that is suppose we have y' LT y. Introduce a new observer moving, as I am, to the left relative to you with the speed v and sharing the common x axis. Because of the left-right symmetry mentioned above, and the fact that we are talking of different y coordinates of one and the same point, the y'' coordinate of the new observer and your y' coordinate will be related in the same way as were your y' coordinate and my y coordinate. Therefore we must have y'' LT y'. Combining this with y' LT y yields that y'' LT y. But the new observer is stationary relative to me, and that means that we have to have $y'' = y$, which conflicts with y'' LT y. If we had started with y' greater than y instead of y' less than y, there would have been a similar conflict. Therefore the only possibility is that $y' = y$, and this means that our relative motion in the common x direction does not affect lengths perpendicular to that direction.

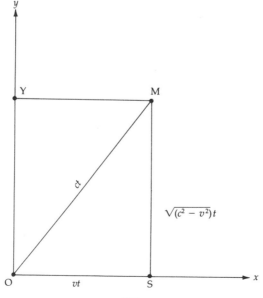

Consider now the light clocks you and I carry. As viewed by me, in one tick the light in my clock travels vertically from O to Y, whereas, again as viewed by me, the light in your clock goes in a sloping direction from O to the moving mirror M. Let t be the time taken for the light to go from O to M according to me. Then the distance OM is ct, and the distance OS is vt. By the Pythagorean theorem, the distance SM (which is the same as the distance OY) is equal to $\sqrt{(c - v^2)}t$. The time taken for light travelling at speed c to traverse this distance is $\sqrt{(c^2 - v^2)}t/c$ or $\sqrt{1 - v^2/c^2}t$. So while the time of a sloping tick is t, that of a vertical tick is $\sqrt{(1 - v^2/c^2)}t$, which is smaller than t. The smaller the intervals between the ticks, the faster the rate of the clock. Thus according to me your clocks go more slowly than mine. But since according to you it is my clock that has the sloping ticks, you will find that my clocks go more slowly than your own by the same factor $\sqrt{(1 - v^2/c^2)}$.

If we were at rest relative to each other, we would expect no slowing down of clock rates, and this is borne out by the fact that if v is zero the slowing factor $\sqrt{(1 - v^2/c^2)}$ has the value unity. What of the extreme—and, indeed, unattainable—case in which our relative speed is the speed of light? Then, since v is here equal to c, the slowing factor has the value zero, so that we would each find that the other's clocks had stopped—though his own clocks were going at their normal rate.

The slowing is strange enough. But it becomes much more so when we notice that the effect is reciprocal. According to your observations (see illustration on page 104), it is not the rays of your light clock but of mine that slope, and from this it follows that you will find it is my light clock that goes the more slowly. Thus each of us says that the other's light clock goes more slowly than his own and by the same factor.

It is interesting to look at the mutual slowing of clock rates in terms of Einsteinian frames of reference with synchronized clocks. In what follows, keep in mind that your clocks are exact replicas of mine and that if they were at relative rest side by side they would all go at the same rate.

Consider first how I go about observing the rate of one of your clocks. Since your clock is stationary relative to you, it is moving relative to me. So it does not remain adjacent to one of my synchronized clocks. I therefore compare its readings with those of two of my synchronized clocks. To illustrate the procedure, it is convenient to represent your clocks by squares and my clocks by circles. A pair of synchronized clocks will be enclosed in a rectangle. Readings of clocks are indicated by numbers inside the circles and squares. Thus the diagram shows that at the start, which we shall think of as time zero, your clock is adjacent to one of my synchronized clocks with both clocks reading zero. Later, your clock has moved so as to be momentarily adjacent to a different one of my synchronized clocks, and at that instant there your clock reads, say, ½ second and my clock reads 1 second. I therefore conclude that your clock is going at only half the rate of mine.

When you observe the rate of one of my clocks, you compare readings of one of my clocks with those of two of your synchronized clocks, as in the diagram on page 106. Since at the left coincidence my clock

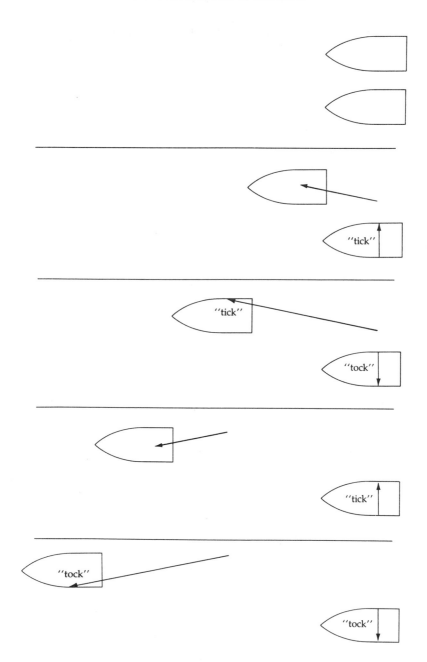

reads ½ second and yours reads 1 second, you conclude that my clock is going at only half the rate of yours. Because in the first case one of your clocks is checked against two of mine and in the second case one of mine against two of yours, there is no conflict. But suppose it had somehow been possible to use only one of your clocks and only one of mine at the same place, as shown on page 107. Then your clock would be reading both 1 second and ½ second at the same instant and the same place—as would mine—which would indeed be a contradiction. But here there is no such direct confrontation. In one case your clock is checked against two of mine, while in the other case my clock is checked against two of yours, and this permits us each to find without contradiction that the other's clocks go more slowly than his own. We will be strongly tempted to say that the mutual slowing down arises because we chose to use such strange clocks. But if we used quartz watches—or, indeed, any trustworthy timing devices—instead of light clocks, exactly the same mutual slowings would have to occur. To see why, suppose you have a quartz watch next to your light clock and that, as determined by you, both clocks go at the same rate. For vividness, imagine the dials of your light clock and your quartz watch superposed, with their seconds hands coinciding as they rotate. When I look at them I must also see the hands coinciding. Therefore, if the light clock seems to be slowed by the relativistic factor the quartz watch must also. Since the same argument holds for all other types of accurate timepieces, we are evidently dealing with a property of time itself. It is called the dilatation of time.

Let us explore further. In the following discussion we shall make use of the fact that the value of a fraction is unaltered if top and bottom are multiplied by the same quantity—for example, $2/3$ and $20/30$ have the same value. Recall that we deduced from Einstein's two principles that, despite our relative motion, you and I find the same value for the speed of light. Lamps in our respective spaceships give off light waves that keep pace. In your spaceship you find the speed of the light waves by measuring the distance the light travels in a given time and then forming the fraction (distance gone)/(time taken). The result must come to c. If I, in my spaceship, do the same, I, too, shall form the fraction (distance gone)/(time taken) and shall also come out with the value c. So far there is no surprise. But now I look at your measurements and observe that, according to me, your clocks are running slow by the relativistic factor $\sqrt{(1 - v^2/c^2)}$. But the fraction (distance gone)/(time taken) has to come out to the same value c as before. So, since I see your "time taken" diminished by the relativistic factor, I must also see your "distance gone" diminished by the identical factor. According to me, therefore, your lengths in the direction of our relative motion are contracted by the

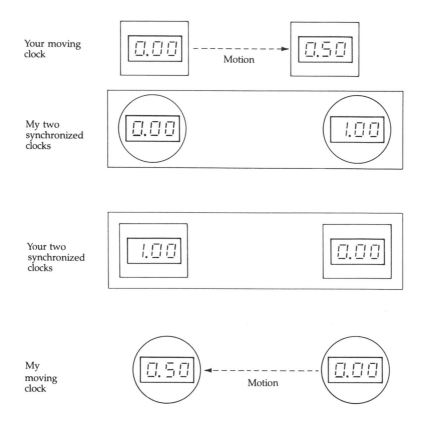

Your moving clock

Motion

My two synchronized clocks

Your two synchronized clocks

My moving clock

Motion

relativistic factor $\sqrt{(1 - v^2/c^2)}$. This reminds us of the FitzGerald-Lorentz contraction—even the amount of the contraction is the same. But there is a difference. If we consider the above situation making use of your observations of my measurements instead of my observations of your measurements, we find that, according to you, mine are the lengths that are contracted. The effect is mutual. Each of us finds the other's lengths in the direction of our relative motion contracted. When FitzGerald and Lorentz and Poincaré spoke of a contraction, they thought of it as arising from motion through the ether. Undoubtedly they silently assumed that someone at rest in the ether would find that moving lengths were contracted but that a moving observer would find that lengths at rest in the ether were expanded compared with his own. And the even greater silence of these scientists about the slowing of clocks shows that in spite of their mathematical equations being the same as Einstein's, the idea of a reciprocal slowing of clocks was foreign to their views.

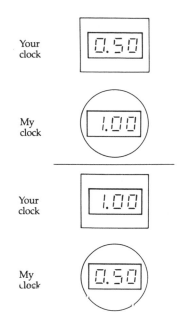

It is appropriate here to recall Fresnel's self-contradictory argument about entrapped ether that led him to his formula for the speed of light in moving glass or other medium. How does his formula arise according to the theory of relativity? Essentially one wishes to combine two speeds, the speed of the medium relative to me and the speed of the light relative to you, with you being at rest relative to the medium. In Newtonian theory one would simply add these two speeds (and thereby obtain a formula in disagreement with experiment). One's first thought concerning the relativistic approach is that the speed of light is c for all observers. But that holds only for the speed of light in a vacuum. Inside the medium the speed is less than c and can therefore be increased. Relativistically one wishes, as before, to add two speeds, the speed of the medium relative to me and the speed of the light relative to you. But I see your measurements being made by means of shrunken rods, slowed clocks, and, of key importance, a foreign simultaneity. When one takes account of these effects one obtains a relativistic formula that yields Fresnel's formula as a first approximation. Little wonder that self-contradiction crept into Fresnel's pre-relativistic argument.

Box 5.4

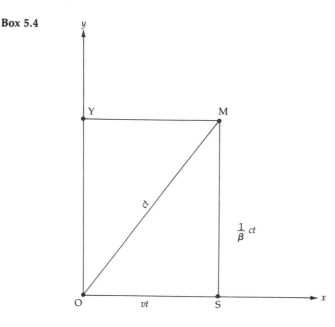

We have seen that because the speed of light is constant, the dilatation of time implies a corresponding contraction of lengths in the direction of our relative motion. According to the argument leading to the contraction, one might expect to have lengths contracted not just in the direction of relative motion but in all directions. As we have also seen, however, the contraction is confined to the direction of motion.

It turns out that for the case of the y direction, and indeed all directions perpendicular to the relative motion, a multiplying factor enters that exactly cancels the contraction of lengths that one would have expected to arise from the dilatation of time. Here are the details.

Suppose you send a beam of light from O in your y direction. If the beam takes time t', as measured by you, to go a distance y' along the y' axis we must have $y' = ct'$. According to me, the beam of light is directed along the sloping line OM, along which it travels with speed c. If t is the time, measured by me, for the journey of the light from O to M, the distance OM is equal to ct. But because OM slopes it is not equal to y, the latter being the length of OY, which is the same as SM. So $y = (1/\beta)ct$, where $(1/\beta) = $ SM/CM $= \sqrt{(1 = v^2/c^2)}$. The time dilatation, according to me, is given by $t' = (1/\beta)t$. The switch from OM to OY, as shown below, thus brings in a factor $1/\beta$ that cancels the $1/\beta$ factor introduced by the dilatation of time, so that we can keep the value c for the speed of light along OM without being in conflict with $y' = y$.

Specifically, from $y' = ct'$, $t' = (1/\beta)t$, and $y = (1/\beta)ct$, we have $y' = ct' = (1/\beta)ct = y$.

A striking confirmation of the special theory of relativity was found by observing electronlike particles called mu mesons, or muons, that are created high in the atmosphere by cosmic rays. Let us first consider the situation relative to the earth. The muons move with a speed very close to that of light, but on the whole they decay into other particles so rapidly that it seemed strange that a large proportion of those that arrived near the top of a mountain should survive long enough to reach the earth. The discrepancy could be understood, however, as soon as one realized that since decay of the muons was a measure of the passage of time, the muons constituted a clock, and hence according to the theory of relativity their decay rates would be found to be slowed down because of their motion. Slowed rates of decay meant longer lifetimes during which the muons could travel, and they thus could travel farther. The experimentally determined numerical results agreed admirably with the relativistic prediction.

Let us now look at the situation from the point of view of an observer moving so as to keep pace with the muons. Since the muons are stationary relative to him, he will not observe a relativistic slowing of their decay rates. But he—and the muons—will see the mountain rush-

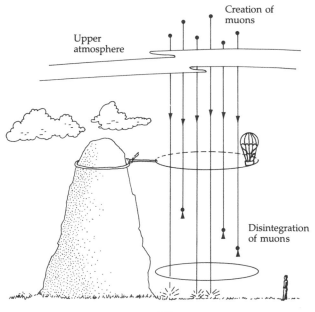

109

ing toward them with almost the speed of light, and therefore relative to them the mountain will be contracted so that the distance between its top and bottom will be much shorter than it was for the observer on the ground. And since, relative to the muons, the factor by which the height of the mountain contracts is the same as that by which, relative to the ground, the time was slowed, the number of muons reaching the level of the base of the mountain will come out to be the same in either frame of reference.

Point of view of muons. The muons travel so fast that the mountain appears 15 times as small as an earth observer sees it.

The muon experiment is a versatile one. It confirms three relativistic predictions: the slowing of clocks, the contraction of lengths, and the similar relativistic behavior of all types of clocks. Note, moreover, that it shows relativity applying to a phenomenon—the decay of muons—that basically belongs neither to mechanics nor to electromagnetism.

Box 5.5

Although the following is not an essential part of relativity, it is of considerable interest. The reader may wish to skim through it first and then look into the details only if eager to do so. For years, scientists believed that because of the FitzGerald-Lorentz contraction, objects moving uniformly relative to us would appear shrunken; in particular, that a sphere moving past us would seem to have the shape of an oblate spheroid—a somewhat flattened sphere. Not until more than half a century after Einstein's original

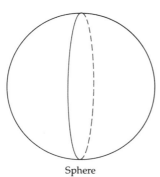

Sphere

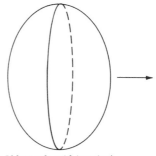

Oblate spheroid (moving)

paper was it realized that in a certain sense that belief was incorrect. The concept was enunciated in 1957 by the American astronomer James Terrell. But because his paper seemed to conflict with long-accepted ideas it was rejected by a succession of distinguished scientific journals and was not published till 1959, by which time the English scientist Roger Penrose had independently developed a special case of the general result.

The basic argument is valid only if, as in astronomical observations, for example, the moving object is so small and so far away that the light rays from it to our eye can all be treated as parallel. Since these restricted viewing conditions may pose problems of visualization, let us consider instead the shapes of shadows of moving objects cast on the horizontal ground by the sun when it is vertically overhead. We may do this because the shape of such a shadow corresponds to the shape of the object that caused it as the object would be seen by us under the restricted viewing conditions, if we were below the object and looking upward at it. We shall use the same letters to label an object and its shadow, the former letters capital and the latter small.

To see the gist of the argument while avoiding the mathematical complexities introduced by considering the appearances of arbitrary shapes in motion, let us consider a moving semitranslucent square ABCD with sides of unit length and with small knobs at the corners to make it easier to see where the corners throw their shadows. Take the square to be vertical and moving to the right, as shown in the diagram. We want to know the shape of its shadow on the ground.

Consider first the Newtonian case and, for the sake of argument, let us pretend that the speed of light is infinite. Then there are no surprises. At a

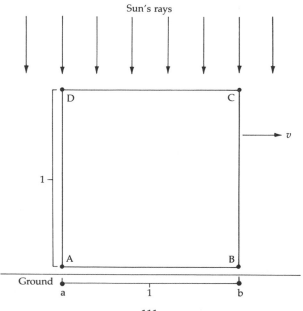

given instant, say, noon, the shadow is a line of unit length with knobs at the ends and coinciding with the position of the side AB at that instant.

Next, while still taking the Newtonian case, let us take into account that light travels with the finite speed c. To find the shape of the shadow at the instant noon, we must consider the pattern of light and shade created by the sunlight that reaches the ground at that instant. As before, there will be a shadow ab of unit length. But there will now be more shadow than that. With the speed of light finite, the sunlight reaching the ground at noon will have been at the CD level a little before noon—to be precise $1/c$ seconds before noon—and at that time the square will have been a distance v/c to the left of its noon position, as shown by the dotted lines. From this one can see that the noon shadow will consist of the part ab of unit length plus a part ad stretching to the left and of length v/c.

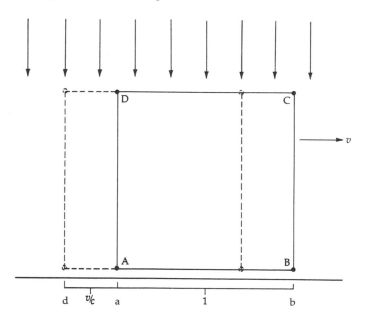

We are now ready to consider the relativistic case. Here the shadow will, as before, be dab except that, because of the FitzGerald-Lorentz contraction, the length of the part ab will be shrunk from unit length to length $\sqrt{(1 - v^2/c^2)}$. Since AD, being perpendicular to the direction of motion, is uncontracted, the distance da remains v/c. And here something extraordinary happens. The Pythagorean theorem suddenly thrusts itself upon us: The sum of the squares of the lengths of da and ab, namely, the sum of the squares of v/c and $\sqrt{(1 - v^2/c^2)}$, has the value unity; so, as shown in the

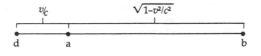

diagram, the shadow dab is the same as we would get if the unit square were stationary relative to us (and thus uncontracted) but in a rotated position. The basic result that a relativistically contracted object appears like a stationary rotated object applies not just to squares but to objects of all shapes. Since the shadow corresponds to the shape we would see under the restricted viewing conditions mentioned above, we can say that under those conditions the visual impression of an object moving uniformly relative to us is not that of a contracted object but of a rotated one. In the case of a moving sphere, the shape will thus seem unaltered with no contraction visible; and, indeed, if there are no markings on the sphere, even the apparent rotation will be invisible.

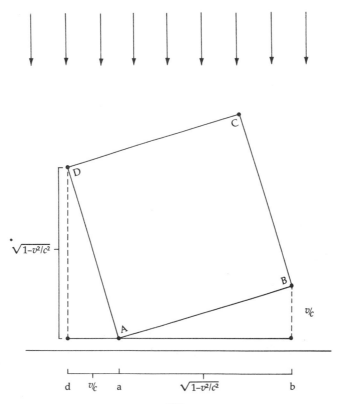

Nevertheless the FitzGerald-Lorentz contraction still exists. When, as above, we look at a square moving relative to us we do not see it as it was at a particular instant; we see the image that is brought to us by light rays that reach our eye at a particular instant, and these rays did not start out from the square all at the same moment.

If we want to envision the moving square all at a particular moment, we have to proceed more circumspectly, using Einstein's hypothetical scaffolding with synchronized clocks at its intersections. Let us imagine that at each clock we station a friend ready to report to us. We ask each of our friends to report whether or not part of the square was at his intersection at a given instant, say noon—this instant being indicated by his local synchronized clock. When the reports are all in, we make a chart showing the location of each reporter and indicating whether or not part of the square was at his intersection at noon. We thus build an image of the moving square as the whole of it was at the single instant noon according to our relativistic simultaneity. The figure that emerges turns out to be an unrotated oblong with sides AD and BC of unit length but sides AB and DC contracted to length $\sqrt{(1 - v^2/c^2)}$, which, of course, is just the amount of the FitzGerald-Lorentz contraction. Indeed the same degree of contraction is found for objects of any shape—and this is true whether the objects are small and distant or not.

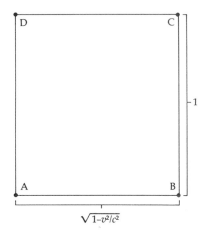

In particular, the image of a sphere comes out to be an oblate spheroid. In this sense the FitzGerald-Lorentz contraction is certainly present. Yet when we merely look at a moving object under the restricted viewing conditions, it is fair to say that the contraction is "invisible."

So far, we have been deriving major relativistic results directly from Einstein's two basic postulates, with minimal use of mathematics. The fact that we have been able to go as far as we have with so little mathematics is a tribute to the power and simplicity of Einstein's two postulates.

No matter how much we remind ourselves that space and time are fundamental, their surprising relativistic behavior is apt to seem amusing rather than important. We tend to forget that space and time dominate physics. Here is an instance of that dominance. It rests on a fact already deduced from Einstein's postulates, namely, that space and time are such that no material object can move as fast as light. Let us consider this fundamental fact in Newtonian terms. It will be convenient here to let you and me be at rest relative to each other and to let all motions mentioned be motions relative to ourselves whether so stated or not. As we know, Newton's second law can be briefly stated in the form "force equals mass times acceleration." The law implies that the greater the mass of a body, the smaller the acceleration produced by a given force acting on it. Mass is thus a measure of inertia—the resistance of bodies to being accelerated.

Suppose a steady force is applied to a small object—say, a stone—that is initially at rest (relative to us). Then according to Newton there will be a steady acceleration—the stone will move steadily faster and faster—and if we continue to apply the force, there should come a time when the stone is moving faster than light. Since we know that according to the theory of relativity that is impossible, we have to wonder what is wrong with this argument.

Basically, the contradiction arises from the mixing of Newtonian and relativistic concepts. In particular, we have assumed, with Newton, that the mass of the stone remains constant. It turns out that in relativity we can retain Newton's second law if we say that mass is relative. As the speed of an object relative to us increases, so, too, does its mass as measured by us, and the mass becomes extremely large as the relative speed nears the speed of light—it would become, in fact, infinite at the speed of light. Specifically, the mathematics shows that, if the mass of an object at rest relative to us is m_0, then its mass m relative to us when moving with speed v relative to us is given by $m = m_0\sqrt{(1 - v^2/c^2)}$. The quantity m_0 is called the rest mass of the body, and m its relative mass.

We can now see why a steady force never gives the stone a speed as great as that of light. When the stone is at rest relative to us, its mass is small, so it will be readily accelerated by the force. The more it is accelerated, the greater its speed relative to us; the greater its speed, the greater its mass; and the greater its mass, the greater its resistance to being accelerated. By the time the stone's relative mass becomes as great as, say, the mass of a mountain, the steady force will be far less effective than it was at the start in increasing the speed of the stone. Because of the stone's ever-increasing resistance to being accelerated, it turns out that an infinite amount of time would be needed for a steady force—or, indeed, any finite force—to accelerate the stone to the speed of light, and that is just another way of saying that the stone could never attain that speed.

Keep in mind in all this that the increase of relative mass with increasing relative speed arises directly from the relativistic changes in our concepts of time and space. Fundamentally it is relativistic time and space that impose the speed limit. The change of mass is only a way of talking about that speed limit in Newtonian terms.

What happens to all the energy conveyed to the stone by the force? The effect of the force is not only to increase the stone's energy but also to increase its speed, and thus its relative mass—its mass when moving relative to us. Given its rest mass—its mass when at rest relative to us— we can find the relative mass for any relative speed. That being the case, we see that the relative mass and the energy imparted that increased the relative speed are different measures of the same thing. So we can think of the relative mass of a particle as a measure of its energy, and Einstein's famous equation $E = mc^2$ does indeed show that relative mass is a measure of energy.

It is worthwhile to indicate how Einstein derived this equation. In a 1905 paper only three pages long he made a relativistic calculation using the electromagnetic field equations, and found that if a body gave off energy of amount L in the form of light, it would lose mass of amount L/c^2. Having found this to be true for energy in the form of light, he said that the fact that the energy ended up in the form of light "obviously makes no difference." And with this daring stroke he once again turned a special case into a universal law. If we denote the energy by E instead of L, we have $E/c^2 = m$, which can be written $E = mc^2$. But that was not the whole story. In 1905 the equality went only from left to right. The equation said that energy has mass.

In 1907 Einstein completed the argument. He considered an object acquiring additional mass by absorbing radiation and argued that one could not reasonably make a distinction between the mass already there and the additional mass acquired, and since the latter was equivalent to energy, so, too, must be the former. Thus even rest mass was equivalent to energy. This meant that every object having mass—even something as seemingly inert as a grain of sand or a feather—is, according to the formula $E = mc^2$, a storehouse of a comparatively enormous amount of energy. For example, a thimbleful of lead contains as much energy as is released in the burning of 100,000 tons of coal. It is energy inherent in mass that powers nuclear bombs.

Among the professors at the Zurich Polytechnic Institute whose classes Einstein cut was the mathematician Hermann Minkowski, who at one time had thought Einstein lazy. Minkowski later became a professor at the outstanding university in Göttingen, Germany, and there, starting in 1907, he showed that the equations of relativity fitted neatly into a structure that he called four-dimensional "space-time." The four-dimensional aspects of relativistic equations had already been significantly developed by Poincaré in the paper that he submitted almost simultaneously with Einstein in 1905, but Minkowski, who went some-what further than Poincaré, is usually given all the credit.

We are already familiar with the basic idea of coordinates. In two dimensions—for example, on a sheet of graph paper—by selecting two perpendicular lines Ox, Oy as coordinate axes with origin O we can specify the positions of points on the paper by means of two coordinates (x, y). In the diagram only the essentials have been drawn, the graph-paper lines having been omitted for the sake of clarity. Through the point P the line PQ is drawn perpendicular to the x axis. If P has the coordinates (x, y), OQ will be of length x and QP of length y. Denote the distance of P from the origin by r. Then because OQP is a right triangle, the Pythagorean theorem gives

$$OP^2 = OQ^2 + QP^2, \text{ or } r^2 = x^2 + y^2.$$

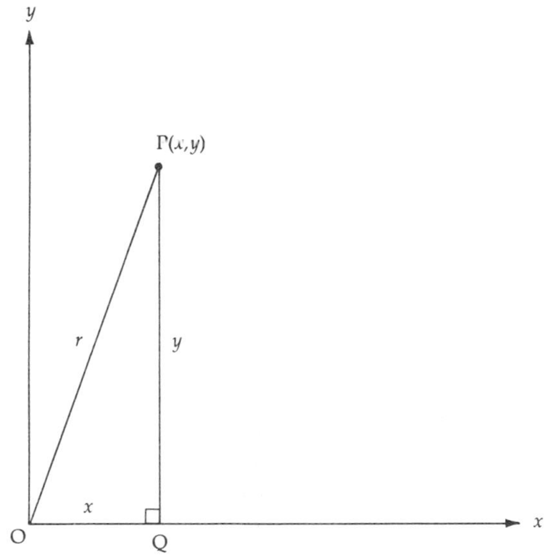

Now introduce a new pair of mutually perpendicular axes with the same origin and at rest in a rotated position relative to the old pair. What happens to the formula for r^2 if we apply a transformation from the old unprimed to the new primed axes? There is a shortcut that leads to the answer, eliminating the need to find the transformation equations that relate the primed and unprimed coordinates. The next diagram essentially reproduces the previous one and shows the same O and P in relation to the rotated axes Ox', Oy'. The line PQ' is perpendicular to the x'

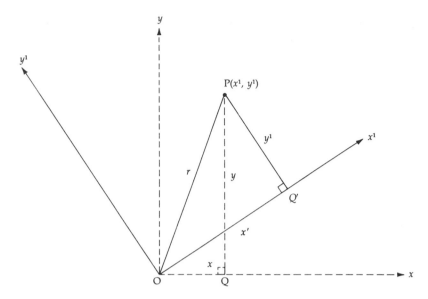

axis. The distance OP, or r, is the same as before, but the x' and y' coordinates of P, given by the distances OQ' and Q'P, differ from the unprimed coordinates of P, given by the distances OQ and QP. Nevertheless, because OQ'P is a right triangle, the Pythagorean theorem gives

$$OP^2 = OQ'^2 + Q'P^2, \text{ or } r^2 = x'^2 + y'^2.$$

Except for the primes, the formula for r^2 in the primed coordinates is exactly the same as the formula in the unprimed coordinates.

In three-dimensional space one can introduce a third coordinate z, the z axis being perpendicular to the other two. By a repeated application of the Pythagorean theorem, it can be shown not only that

$$r^2 = x^2 + y^2 + z^2,$$

but also that when one applies a transformation to a primed mutually perpendicular triplet of axes with the same origin, this becomes

$$r^2 = x'^2 + y'^2 + z'^2.$$

If we look at the Lorentz transformation, we see that x and t are inextricably mingled, and this alone is a powerful reason for suspecting that time will mingle geometrically with space. An algebraic calculation shows that under a Lorentz transformation the quantity s given by

$$s^2 = x^2 + y^2 + z^2 - c^2t^2$$

is such that we also have

$$s^2 = x'^2 + y'^2 + z'^2 - c^2t'^2,$$

118

and, in spite of the c^2 and the minus, this is so strongly reminiscent of the formulas for r^2 in ordinary two-dimensional and three-dimensional geometry that it makes a four-dimensional interpretation inevitable, with time linked to space in full partnership to form a four-dimensional world. In this four-dimensional Minkowski world the quantity s, which is analogous to the distance between two points, is called the interval between two events. [A slightly more complicated formula holds when neither one of the events has its space-time coordinates (x, y, z, t) all equal to zero.] Just as the formula for the distance r retains its form when we make a rotatory transformation of axes in space, the expression for the interval s retains its form when we make a Lorentz transformation in space-time. From all this, the Lorentz transformation emerges as being closely analogous to a transition to rotated coordinate axes.

As an indication of the importance of the interval, suppose that we are in our spaceships travelling uniformly relative to each other, as usual, at a constant high speed, and you decide to play a game of chess. Your opening move, P–Q4, involves two events: picking up the queen pawn and placing it two squares forward on the Q4 square. For you these events are separated by, say, 6 centimeters of distance and, say, 1 second of time. Because of the fast motion of your spaceship relative to me, the two events, according to my measurements, would be separated by, say, 1000 kilometers of distance; and because I see your clocks going at a slower rate than mine, the two events, according to my measurements, would be separated by a little more than 1 second of time— approximately 1.0000056 seconds. As expected, we differ as to both the spatial and the temporal separations of the two events. But despite our disagreements, when we each calculate the interval between the two events according to our own space and time measurements, we come

CHESS GAME

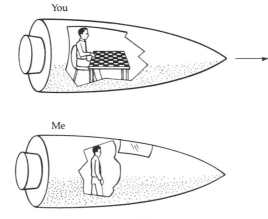

119

out with the same value. After so many disagreements between our measurements, it is refreshing to find something on which we—and everyone else in unaccelerated motion—agree. Clearly, the interval will prove useful in picturing what goes on in the four-dimensional world of the special theory of relativity.

Nobody can fully visualize four dimensions. For easy visualization of key aspects of space-time, it is customary to suppress two of the three spatial dimensions by only considering the region in which y and z are zero—namely, the x axis—and thus to consider only two coordinates, the distance x and the time t. It is convenient to replace the time coordinate t by the distance ct travelled by light in time t, the advantage being that the coordinates x and ct both represent distances. Let us assume that I measure these coordinates in my reference frame, and that I am seated comfortably at the place where $x = 0$. As is customary, let us simplify by thinking of me (and other persons and events) as having negligible size. It may seem that since I am resting where $x = 0$, I can be represented by the point event O in the Minkowski diagram. But time does not stand still. The coordinate ct steadily increases whether we like it or not. When I am resting where $x = 0$, I am therefore not represented by a dot at event O but by a segment of the vertical line constituting the ct axis: the older I get, the higher my representative point on this segment of the ct axis. The segment is called my world line. What of you as viewed by me? Assume for simplicity that you start next to me at the central event O of the diagram, namely, at $x = 0$ when $ct = 0$, and move along my x axis at constant velocity. Since your x coordinate steadily increases with time, your world line will be like the line shown in the diagram. In fact, the world line of any particle acted on by no forces, and thus moving relative to me with constant velocity, will be a

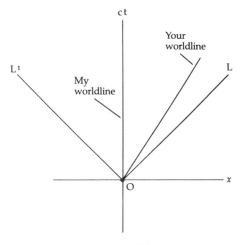

straight line in the Minkowski world, though not, in general, passing through event O and not, in general, confined to the region in which y and z are zero. Because particles moving with constant velocity have straight world lines in Minkowski space-time, Newton's first law, which says that free particles move with constant velocity, can be summed up by saying that the world lines of free particles are straight.

When we try to draw space-and-time slices of the Minkowski world on paper and try to treat the interval s as we treat distance r in ordinary geometry, it is impossible to avoid severe distortion. This is shown strikingly by the two lines OL, OL' making 45-degree angles with the axes. The horizontal coordinate x of any point on the line OL is equal to the vertical coordinate ct of that point, so we have $x = ct$ along OL. But if a light ray starting at $x = 0$ when $t = 0$ is shone in the positive x direction, its tip will travel in that direction with the speed of light, and since for uniform motion distance gone is equal to speed multiplied by time taken, the tip will move in such a way that $x = ct$. Thus its world line will be OL. The tip of a ray sent in the opposite direction would have OL' as its world line. But if x is equal to ct (and y and z are kept equal to zero), it follows at once from the formula for s^2 that $s = 0$. So the interval from event O to any other event on OL, and similarly to any event on OL', is zero—and this is the only measure along OL or OL' on which all uniformly moving observers agree. There is nothing to be done about the disparity between the zero interval and the nonzero distance in the diagram except to keep it in mind. After a while one learns to live with it.

As we have seen, according to the theory of relativity, material objects cannot move as fast as light. For such objects starting out from event O, the quantity s^2 defined above turns out to be negative. To avoid the awkwardness, one redefines s^2 as the negative of the expression given above while still calling the corresponding s the interval. This redefined interval has an interesting significance. Consider the portion OP of the straight world line of an unaccelerated observer. If, as is customary, he remains at the place where his x, y, and z are all zero, his world line will lie along his ct axis. With x, y, and z all zero, the new formula for s^2 reduces to $s^2 = c^2t^2$, so that we can take $s = ct$. This says that, apart from the factor c, the interval s from the event O to event P gives the lapse of time from O to P as measured by an accurate watch carried by the observer and therefore also as measured by the amount by which the observer has aged in the passage from event O to event P.

For events neither of which is the event O, one has to use a slightly more complicated formula for s^2, but it turns out that analogous remarks still hold. If the observer is accelerated, his world line will be curved, and we would then need formulas from the calculus to find intervals between events on the curved world line as measured along that world line. But the same result emerges as before: apart from the factor c, the interval gives the time between the events as measured by the observer's watch, and thus by the amount of his aging. Because of this, the interval divided by c is called proper time—not in any moral sense of propriety but in the rarer sense of personal belonging—as in the word property.

Let us now briefly look at the Minkowski world with only z held equal to zero so that we take account of two spatial dimensions and one temporal dimension of four-dimensional space-time. We now have a three-dimensional picture, and in it the former lines OL and OL' expand into the cone formed by all lines through the event O that make a 45-degree angle with the ct axis, a cone with a future part and a past. It is called a light cone.

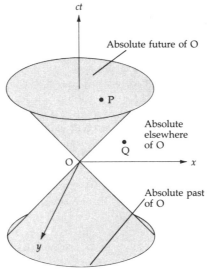

There is an observer whose world line lies along the ct axis. The greater the speed of a particle relative to that observer, the farther the particle moves from him in a given time, and therefore the greater the angle its world line makes with the ct axis. Since no particle can travel faster than light, no particle can travel farther than light in a given period of time, and the world lines through event O of physical particles must therefore lie within or on the light cone. Any event such as P within or on the future part of the cone can be reached from event O without exceeding the speed limit, and it turns out that all observers will agree that the event occurred later than event O. For this reason the region within and on the future part of the cone is called the absolute future of event O. For analogous reasons, the region within or on the past part of the cone is called the absolute past of event O. An event outside the light cone—for example, Q—cannot be reached from event

O by anything moving with a speed less than or equal to the speed of light. Thus event Q cannot be caused by event O, and since a cause cannot precede its effect, this neatly dovetails with the fact that for some observers O will be later than Q, while for others it will be earlier than Q. The region outside the light cone is called the absolute elsewhere of event O.

Newtonian space and time can be regarded as the special case of Minkowski space-time in which the speed of light is infinite. In this case the light cone flattens out so that the absolute elsewhere region vanishes, leaving only absolute future and absolute past separated by a fleeting absolute present, as one would expect of Newton's absolute time.

Some aspects of relativity cause particular puzzlement—for example, the fact that you and I each observe the other's meter rods to be shortened. Yet there are everyday mutual contractions that we accept quite readily. If two men of equal stature walk away from each other, each sees the other getting smaller than himself, yet nobody seems upset about it. We have already remarked that the Lorentz transformation is analogous to a rotation of coordinate axes. When I use my own meter rods, they are at rest relative to me so that I view them squarely and measure them at their actual lengths. When I measure your meter rods, it is as if I were viewing them at an angle and thus foreshortened. Correspondingly, you view your own rods squarely but mine as if obliquely and thus foreshortened. The situation with the reciprocal slowing of clocks is analogous but harder to follow because we are less accustomed to thinking of time geometrically.

Consider, now, a famous relativistic problem—the problem of the twins. One twin stays at home on the earth; the other one travels. The travelling twin leaves the earth on a spaceship, travels for, say, a year at extremely high speed, then reverses direction, travels for another year, and arrives back on the earth. In the course of his journey the traveller has aged two years. But at his reunion with his stay-at-home brother, he finds that his stay-at-home brother has aged by, say, fifty years and is thus now forty-eight years older than he is.

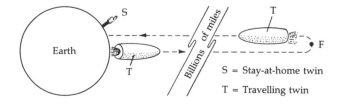

S = Stay-at-home twin
T = Travelling twin

Let us consider this phenomenon first in terms of the slowing down of the rates of clocks because of relative motion. The travelling twin is a sort of clock; so, too, is the stay-at-home twin. If we doubt this fact, we can have the twins carry watches capable of counting the passing years, and the watches will confirm their disparate agings. As viewed by the stay-at-home twin, the clocks and aging of the travelling twin will seem to go more slowly than his own.

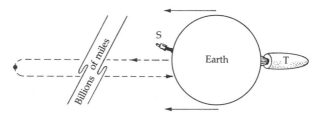

But some people have objected that the slowing of clocks is a reciprocal relation. They point out that each observer finds that the other's clocks go more slowly than his own. Suppose, the argument goes, that we look at the situation relative to the travelling twin. Then the travelling twin will seem to stay put and the stay-at-home twin will seem to do the travelling. Therefore, people argue, when the twins meet again it should be the stay-at-home twin who has aged by two years and the traveller who has aged by fifty. It was puzzling enough to have the traveller find his brother much older than himself when he returned. But even in relativity it is quite impossible for one twin to find himself both older and younger than the other when they meet again.

Actually the twins cannot legitimately be treated reciprocally, as in the preceding paragraph. There is a crucial difference between them that is best seen by making the reversal of direction of the spaceship after one year an abrupt one—say, one taking 30 seconds. Then the traveller would experience a deceleration force of about a million times the pull of earth's gravity, and he would at once be squashed flat against the wall of his spaceship. But when we look at the situation relative to the travelling twin with the stay-at-home twin now the apparent traveller, the stay-at-home twin would nonetheless experience no such lethal force, while the traveller still would.

Let us now consider the problem from the point of view of space-time. First we remark that the travelling twin does not age more slowly than his brother. The two twins age at exactly the same rate. If we used twins aging at different rates, there would be no need to have one of them travel. They would age differently even sitting side by side. If we gave the travelling twin a watch adjusted to go more slowly than the one we gave the stay-at-home twin we would, once more, not need to introduce travel.

What, then, explains the fact that at the reunion the travelling twin has aged less than his stay-at-home brother? Consider an analogy. One man drives by car directly from A to C while another drives from A to B and then from B to C. On consulting their odometers they find that although they both started at A and ended at C, they travelled different distances, and no one is in the least surprised or upset.

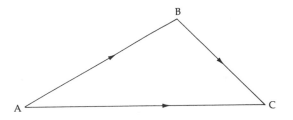

Let us now draw the world lines of the twins in space-time, starting with them together at the moment of takeoff and ending with them reunited when the spaceship lands. The "odometers" here are the aging twins themselves, or the clocks they carry with them that tick off their proper times and thus their aging. The stay-at-home twin has the world line AC, while the travelling twin has the world line ABC. And there should be no surprise that the proper time for AC is different from that for ABC.

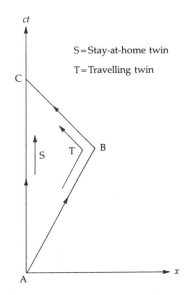

There is one surprising thing, though. The travelling twin came back younger than his brother, but noting that ABC is longer than AC we might well expect him to be the older rather than the younger at the reunion. But we are forgetting the inevitable distortions when we draw Minkowski diagrams on ordinary paper—recall that proper times along the light cone are zero. It turns out that in space-time ABC is indeed a shorter interval than AC.

The relativistic prediction has actually been tested experimentally in a somewhat more general situation in which gravitation also enters. In bare outline the procedure was this: Extremely accurate atomic clocks were used, one of them remaining on the ground while another was taken on a jet plane on a trip around the world. When the travelling clock was reunited with its stay-at-home twin, it had recorded a smaller lapse of time by an amount agreeing with the relativistic prediction.

The nature of great science is strikingly illustrated by the following incident involving experiment. We have seen how Einstein built his aesthetically inspired special theory of relativity on two daringly simple principles. In 1905 he derived relativistic equations predicting the way electrons would move in an electromagnetic field, his prediction agreeing with one that Lorentz had made in 1904. In 1906 the German experimenter Walter Kaufmann published newly obtained results of his experiments designed to test this prediction. Right at the start of his paper he said, "The measurement results are not compatible with the Lorentz-Einstein fundamental assumptions." He went on to say that his measurements favored two other theories. Lorentz, shaken by the adverse verdict of experiment, was ready to discard his hard-won equations. Einstein, however, remained unperturbed, saying, "In my opinion both [of the other] theories have a rather small probability [of being right] because their fundamental assumptions concerning the mass of moving electrons are not explainable in terms of theoretical systems that embrace a greater complex of phenomena." We see Einstein here placing more confidence in his sense of beauty than in the results of experiment. As for the Kaufmann results, they were later found to be erroneous, and more careful experiments agreed with the Lorentz-Einstein equations.

Back in 1905, when Einstein propounded the special theory of relativity, experiments using extremely high-speed particles were a rarity. Nowadays they are commonplace, and they have put Einstein's theory to searching tests that it has passed brilliantly. His theory pervades modern laboratory physics. Its triumphs therein have been numerous and often striking. Yet its greater triumphs have been outside the laboratory, in the study of the theorist. For by its example it inspired daring advances. And it led the British theorist Paul Dirac to a relativistic quantum equation for the electron that is justly praised for both its beauty and its agreement with experiment. Above all, the special theory was, for Einstein, an essential stepping stone to his general theory of relativity.

Let us end this chapter by looking back at the strange adventure that has befallen the Newtonian principle of relativity so far. For processes belonging to the Newtonian sciences of mechanics, the principle was an immediate consequence of Newton's laws, even though those laws were set in absolute space. With the advent of the wave theory of light and its ether, the principle of relativity seemed doomed: Although it still held for Newtonian mechanics, it should not hold for optics. The unexpected failure of optical and other experiments to show any effect of uniform motion through the ether led Poincaré and Einstein to think of the principle of relativity as a basic physical law. And Einstein, by adjoining his second postulate and giving up absolute simultaneity, derived the Lorentz transformation not only as characterizing relative motion in electromagnetic theory but in all branches of physics. However, while Maxwell's electromagnetic field equations fit beautifully into the relativistic framework, Newtonian mechanics does not. If we apply a Lorentz transformation to Newtonian equations, terms involving the relative velocity v enter the transformed equations, and this shows that they do not conform to the principle of relativity. With the Lorentz transformation applying to all of physics, we cannot go back to Galilean transformations for the Newtonian equations. But neither can we evade the principle of relativity—it is a basic postulate of the theory.

The logic of the structure of special relativity therefore forces us to alter Newtonian mechanics to conform to the principle of relativity when we use relativistic notions of time and space instead of Newtonian ones; that is, when we use Lorentz instead of Galilean transformations. Out of this come, among other things, the increase of relative mass with increasing relative speed, and the equivalence of mass and energy summed up in the fateful equation $E = mc^2$. And, above all, there comes a new understanding of the behavior of space and time.

There was one major misfit: Newton's inverse-square law of gravitation with its instantaneous action at a distance. For in the theory of relativity, the prohibition against speeds greater than the speed of light applies to any form of energy. If gravitation really acted at a distance instantaneously, the whole structure of the theory of relativity would be undermined, since by a flick of the wrist one could displace the mass of one's hand and thus, however minutely, alter the gravitational forces not just in one's vicinity but instantaneously everywhere. With such instantaneous signals one could synchronize all clocks simultaneously throughout the universe, thus making simultaneity absolute—in which case it could not be relative. What happened to Newton's law of gravity is a matter that belongs to Einstein's general theory of relativity, which is taken up in the next chapter.

The General Theory
of Relativity

CHAPTER 6

I$\text{T TOOK EINSTEIN TEN YEARS}$—$\text{FROM}$ 1905 to 1915—to go from his special theory of relativity to his masterpiece, the general theory of relativity. In 1912, with the general theory still in the making, he wrote to a friend: "In all my life I have never before labored so hard. . . . Compared with this problem, the original theory of relativity is child's play."

In spite of his scientific accomplishments, Einstein had to wait in Bern at the patent office until 1909 before a significant academic position came to him. In that year he was appointed assistant professor at the University of Zurich, adjacent to the Polytechnic Institute. In 1911 he became a professor at the German University in Prague. In 1912 he came back to Zurich as a professor at the Polytechnic Institute, whose entrance examination he had failed some seventeen years before. He barely had a chance to settle down comfortably in his old academic haunts when he received an offer he could not resist, and as a result, in 1914, on the eve of World War I, he went to Berlin as a salaried member of the renowned Prussian Academy of Sciences with the title of Professor at the University of Berlin, and as the future director of the planned Kaiser Wilhelm Institute for physics—and all this with the freedom to devote himself, if he so desired, solely to his research.

Einstein's general theory of relativity had its origin in an aesthetic dissatisfaction. We can understand why rest and uniform motion are relative if we think of space as having no fixed milestones or other such markers against which to measure absolute motion. But a nagging problem remains. The lack of markers in space explains too much: it implies not just that rest and uniform motion should be relative, but that uniform acceleration, and, indeed, all motion, should be relative, and this seems definitely not to be the case.

We do not need a turbulent plane flight to convince us. A ride on a roller coaster offers just as persuasive evidence that, without looking outside, we can detect the acceleration of a vehicle in which we are travelling, and therefore that acceleration is absolute. And, to take something more commonplace, we all know that when standing in a crowded train we can relax when the motion is smooth but have to hold on when the vehicle jerks.

Back in Newton's day, Berkeley and Leibnitz had sharply criticized the idea of absolute motion. And in the nineteenth century the scientist and philosopher Ernst Mach had argued cogently that the effect of non-uniform motion must arise not from motion relative to Newton's absolute space but from motion relative to the so-called fixed stars and all other matter in the universe. Both Berkeley and Mach were arguing that there is no absolute motion and that all motion is relative. But Einstein had something more specifically physical in mind.

Using a new space and time, he had already shown in his special theory of relativity that uniform motion is relative. To him it seemed intolerably ugly to have uniform motion relative but other types of motion not: to have a spaceship speed up from 5 to 7 kilometers per second in one second, and to be unable to detect either the 5 kilometers per second or the 7 kilometers per second and yet to be able to detect the acceleration of 2 kilometers per second per second. Why should something as fundamental as the principle of relativity apply only to rest and uniform motion? All types of motion should be relative, or else all should be absolute. And since according to Einstein's own special theory of relativity rest and uniform motion were already relative, all motion would simply have to be relative—not just in a philosophical sense, such as that of Berkeley and Mach, but in a physical sense, analogous to that pertaining to uniform motion in the special principle of relativity. So Einstein enunciated a general principle of relativity, according to which if a vehicle is moving in any way at all, there is no interior effect that will let us detect any aspect of its motion in an absolute sense. Or, to use his own words, "The laws of physics must be of such a nature that they apply to systems of reference in any type of motion."

This principle seems to be manifestly contrary to the facts, as witness the airplane, the roller coaster, and the jerking train—and, for good measure, seasickness. A lesser man would at once have dismissed this general principle of relativity as a pleasing fantasy that unfortunately conflicted with elementary facts of everyday experience. But Einstein, trusting his aesthetic sense, retained his general principle of relativity and looked at the antagonistic facts with a fresh eye. To his delight he saw that instead of disproving the general principle of relativity, they actually supported it. That realization ranks as one of the key advances in the history of physics.

Before telling how Einstein found support for his general principle of relativity, let us pause a moment to recall the harsh opposition to the idea of a moving earth. According to the general principle of relativity, all reference frames, no matter what their relative motions, are on a par for the expression of physical laws. Thus we are free to use a reference frame relative to which the earth is at rest, and equally free to use a different reference frame relative to which the earth moves in orbit around a stationary sun. In this sense, Ptolemy and Copernicus were both right, and the bitter conflict between their adherents would seem to have been in vain. But in a profound sense the conflict was necessary, since it led to the acceptance of the difficult concept of a moving earth and thereby prepared the way for Newton and, in an unexpected twist, for Einstein.

Like others, Einstein had tried to treat gravitation in terms of his 1905 theory of relativity. It was the obvious thing to do, and it would automatically replace instantaneous gravitational action at a distance by propagation of gravitational action with speed c. But in the course of the calculations Einstein concluded that, according to this approach, the rate of fall of an object would depend on its horizontal speed—in conflict, for example, with Galileo's neat theory of the motion of cannonballs. That led Einstein to look afresh at Galileo's discovery that falling bodies, whatever their mass, fall with the same acceleration. Let Einstein tell us in his own words what happened next. It will give us an additional instance of his extraordinary intuition. "This law [of falling bodies]," he said, "which may also be formulated as the law of the equality of inertial and gravitational mass, was now brought home to me in all its significance. I was in the highest degree amazed at its existence and guessed that in it must lie the key to a deeper understanding of inertia and gravitation." In all the years since Galileo and Newton, nobody before had realized that there was something here worth noticing. Here is Einstein's argument. He adumbrated it in 1907, returned to it in 1911, and used it as a basic principle thereafter.

Imagine a small laboratory so far removed from other objects that we can think of it as unaffected by gravity. Let an angel give it an acceleration g equal to the acceleration that the earth imparts to falling bodies. (For a different acceleration the earth would have to be replaced by a different body—one that produced the appropriate acceleration in falling objects.) Compare what happens inside the laboratory with what happens inside a similarly equipped small laboratory on the earth. Let us refer to the first laboratory as the sky laboratory and to the second as the earth laboratory. As usual, air resistance will be ignored, and we shall assume that because of the small size of the earth laboratory, we may take the gravitational field inside it to be everywhere the same.

Suppose, first, that the angel is not accelerating the sky laboratory. The experimenter inside the laboratory holds two spheres, one massive and one light, and suddenly lets go of them. Then, by Newton's first law,

they will share any uniform motion the laboratory may have and will therefore remain stationary relative to it. If, however, the angel had been accelerating the sky laboratory, what would happen when the experimenter let go of the spheres? Those spheres, being unaccelerated while the laboratory is accelerated "upward," would, relative to the laboratory, be accelerated "downward" with acceleration g. This reminds us of Galileo's discovery that all objects fall to the earth with the same acceleration, g, and we realize that the above experiment performed in the sky laboratory gives the same result as a corresponding experiment performed in the earth laboratory. The laboratories should not be large. If the earth laboratory were enormous, say 5000 kilometers (3000 miles) wide, we would notice that dropped bodies did not have parallel paths but fell radially toward the center of the earth, and that plumb lines at different locations were likewise not parallel. It can be shown, though, that for small laboratories, any mechanical experiment performed in the sky laboratory will give the same results as the corresponding experiment performed in the earth laboratory. Thus, as far as mechanical phenomena are concerned, there is a sort of equivalence between the two laboratories. This realization was no mean feat.

But Einstein could not stop here. It would be most inartistic to have so fundamental an equivalence apply only to mechanics and not to all of physics. God would not have made the universe in that way.

So, by a stroke of genius, Einstein broadened the partial equivalence into a total equivalence, saying that every experiment in the sky laboratory, whether mechanical or not, will yield the same results as the

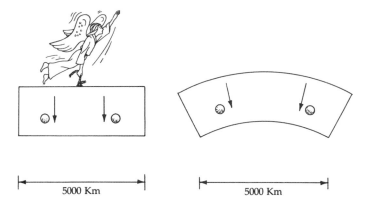

corresponding experiment in the earth laboratory. He called this the principle of equivalence.

Let us look at this equivalence in another way. Suppose we are in an elevator in a skyscraper. While the elevator is stationary, the gravitation of the earth is clearly evident inside it; for example, dropped objects all fall with the same acceleration g. Suppose, now, that the supporting cable snaps and the safety devices fail, so that the elevator plummets. The elevator will fall with the same acceleration as other objects, so objects "dropped" inside the elevator will remain at rest relative to it and will thus seem to an observer inside the falling elevator to float weightlessly. The observer, too, will seem to himself to float weightlessly. Indeed, all phenomena inside the falling elevator will occur relative to it as if it were stationary and the earth's gravitation had been switched off in its vicinity. This cancellation of gravitation by free-fall came vividly to Einstein at a very early stage in his investigations. It reminded him of the electric field produced by a moving magnet—the field vanishes for an observer at rest relative to the magnet. He concluded that, in small regions though not in large, gravitation has only a relative existence.

The principle of equivalence links uniform acceleration and uniform gravitation. But the gravitational fields of stars, planets, and the like depart from uniformity in a crucial way. For example, at the poles of the earth, or indeed at any two antipodal points on its surface, its gravitation acts in opposite directions. Such a gravitational field cannot be mimicked inside an enormous sky laboratory by accelerating the sky laboratory. But, as Einstein well realized, if we limit ourselves to small nonrotating laboratories and small intervals of time, the equivalence of acceleration and gravitation holds to a high degree of accuracy.

133

Box 6.1

In his principle of equivalence Einstein proposed an intimate link between acceleration and gravitation. In particular he pointed out that mechanical effects inside a small uniformly accelerated laboratory would be duplicated in a small unaccelerated laboratory in a uniform gravitational field, and he extended the equivalence to include not merely mechanics but all of physics.

In order to have all motion relative, he needed the equivalence to hold for all types of motion. The case of uniform acceleration posed no problem but what of the complex accelerational effects inside an airplane in turbulent flight? Suppose, first, for the sake of simplicity, that the acceleration, while varying in time, remains all the while in a fixed "upward" direction. Then free objects inside the airplane will all have a common varying downward acceleration relative to the airplane, so the acceleration effects in the airplane will be matched in an earth laboratory if the intensity of the gravitational field in the latter is varied to correspond in time with the varying acceleration instead of being held constant. Admittedly, we would have difficulty in physically producing the needed varying gravitational field inside the earth laboratory without cheating by accelerating either the earth or auxiliary massive gravitating objects. In spite of this, we would be apt to retain a comfortable feeling that in this particular case the equivalence still holds between acceleration and gravitation.

A Space Station
Rotating Uniformly

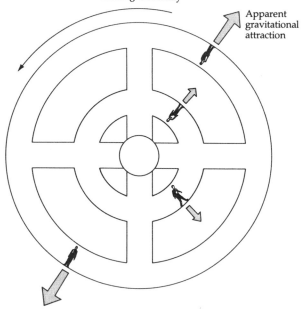

Apparent
gravitational
attraction

If our sky laboratory is the rotating wheel of the space station from the movie *2001: A Space Odyssey*, then all objects that seem to be at rest with respect to the station experience an apparent outward gravitational pull whose magnitude is proportional to the distance from the center of rotation and whose direction is directly away from that center.

But when the acceleration of the airplane or the sky laboratory changes direction there will be rotation, and rotation presents a problem. For example, imagine the sky laboratory no longer accelerated "upward" but, instead, rotating at a constant rate about an axis within it. Consider extremely small laboratories fixed at various places within the rotating laboratory. Each of these tiny laboratories will move in a circle with constant speed as viewed from a non-rotating laboratory. In discussing Newton's orbiting cannonball we learned that a point moving in a circle with constant speed is accelerated toward the center. It turns out that for a fixed angular speed the acceleration is greater the greater the radius. Thus the various tiny laboratories will be accelerated toward the axis of rotation, the acceleration being zero on the axis and increasing as the distance from the axis increases. By applying the principle of equivalence to the tiny laboratories, we see that the rotating laboratory is equivalent to a gravitational field that points radially outward from an axis and increases with increasing distance from the axis. In Newton's theory, but not necessarily in Einstein's, there is no arrangement of external gravitating objects that could produce such a gravitational field.

Feasible or not, such a field of force, precisely because it would have to mimic the accelerational effects in the rotating laboratory, would have to accelerate any free particle at a given place by the same amount no matter what the mass or chemical composition of the particle. It would thus share with actual gravitational fields the key property that we have spoken of as the equality of inertial and gravitational mass. Elaborating on Mach's idea that a particle acquires inertia because of an interaction with all the other matter in the universe, Einstein argued that that interaction must be gravitational. He could then regard even the most nonuniform acceleration effects in a small laboratory as equivalent to gravitational effects and could thus have all such motion relative.

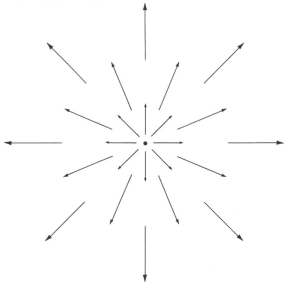

These arrows show a field of force always pointing away from the center of rotation and with a magnitude proportional to the distance from the center.

135

With all motion relative, the rotational acceleration effects in a laboratory should be the same whether it is the laboratory or the rest of the universe that is doing the rotating. One would expect, therefore, that if a laboratory were inside a huge, massive, rotating, spherical shell—a substitute for the rotating universe—the gravitational field would give rise inside the laboratory to the sorts of effects that are found inside a rotating laboratory with the rest of the universe not rotating. According to the general theory of relativity, effects of this sort do indeed occur. However, the postulate of Mach and Einstein that all inertia arises from an interaction with the rest of the universe is not wholly incorporated into the general theory of relativity. For example, in that theory it is possible for a single mass in an otherwise empty universe to have a gravitational field, whereas according to the Mach-Einstein postulate, a lone particle in an otherwise empty universe ought not to possess mass, and thus ought not to have a gravitational field. Despite this, the Mach-Einstein postulate was of key importance to Einstein in his development of the general theory of relativity.

The fecundity of the principle of equivalence was extraordinary. Right away, Einstein could demonstrate that constant acceleration is not absolute but relative. For example, if all free particles in a laboratory fall with constant acceleration g, an observer in the laboratory cannot conclude that it is accelerated upward with constant acceleration g. It might be at rest on the earth instead. Indeed, no experimental effects in the sky laboratory could unequivocally be assigned to acceleration: They could equally well arise from uniform gravitation, or from infinitely many appropriate mixtures of acceleration and gravitation.

With the principle of equivalence linking gravitation to acceleration and showing that acceleration is relative, Einstein saw again that an acceptable theory of gravitation would not fit into the special theory of relativity, in which acceleration was absolute. This in itself was a major indication to Einstein of the path he should follow to arrive at a more general theory that would embrace gravitation.

Also, before that general theory could be formulated, Einstein was able to use the principle of equivalence to discover facts about gravitation by considering experiments in the sky laboratory and interpreting them in terms of the earth laboratory.

Let us look at some examples. The experimenter in the accelerating sky laboratory suspends a mass by means of a spring, and the experimenter in the earth laboratory does likewise. There is no gravitation in the sky laboratory. The spring becomes extended because of the reluctance of the mass to be accelerated, so the extension of the spring measures the inertial mass of the body. In the earth laboratory there is no acceleration, and the extension of the spring measures the gravitational mass of the body. Since, by the principle of equivalence, corresponding experiments must give the same results, the springs will be extended by equal amounts, from which it follows that inertial mass equals gravitational mass. That is something we already knew. But now let the bodies

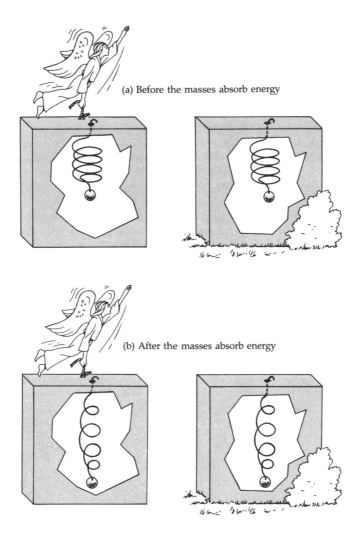

(a) Before the masses absorb energy

(b) After the masses absorb energy

in the two laboratories absorb equal amounts of energy and thus increase their masses by equal amounts in accordance with Einstein's formula $E = mc^2$. Then the springs will be extended further by equal amounts, and from this one obtains the beautiful unifying result that, like other mass, the inertial mass of energy is equal to its gravitational mass.

In another exploration Einstein imagined the following experiment being performed in both the sky laboratory and the earth laboratory. In each laboratory take two accurate clocks that go at the same rate, and fix one on the floor and the other on the ceiling. Our task is to

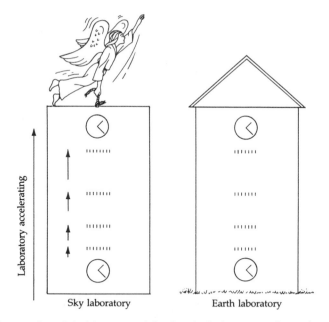

Sky laboratory   Earth laboratory

In an accelerated sky laboratory, and therefore also in the corresponding earth laboratory, the frequence of arrival of light pulses from the clocks below is lower than the ticking rate of the upper clocks even though all the clocks go at the same rate. This is because, in the sky laboratory, the upper clock is receding from the travelling pulses at a constantly increasing rate. The observed slowing is called the gravitational red shift.

compare the rates of the clocks. Since I have just said that they go at the same rate, there seems to be no point to the experiment, especially because, throughout the argument, I shall be insisting that the two clocks do indeed go at the same rate. Let us consider it nevertheless.

The experimenter climbs to the ceiling and looks at the floor clock. To compare the rates of the two clocks, he obviously has to see them both, and in analyzing the situation it is necessary to take into account the light that travels from the floor clock to the ceiling. For the sake of simplicity, let the floor clock send out a brief pulse of light at each tick. Because of the acceleration of the sky laboratory, the ceiling will be receding faster and faster from the rising light pulses, so each pulse will take a longer time than its predecessor to travel from floor to ceiling. The pulses will therefore arrive at a rate slower than the one-per-tick rate at which they started. As a result the experimenter at the ceiling of the sky laboratory will see with his own eyes that the floor clock is going at a slower rate than the ceiling clock—even though, as I have stressed, both are going at the same rate. Note that there is no reciprocity here akin to the mutual slowing of clocks familiar to us from the special theory of relativity. If the experimenter looks at the ceiling clock from the floor, he will be an approaching target for the light signals from the ceiling clock, and he will therefore see the ceiling clock going faster than the floor clock, not slower.

Let us now consider the same experiment performed in the earth laboratory. By the principle of equivalence, the result must be the same as in the sky laboratory: The experimenter next to the ceiling clock must see the floor clock going at a slower rate. But here there is no acceleration to account for the result. Instead there is gravitation. So Einstein concluded that if two clocks going at the same rate are placed with one closer to the earth than the other, the one closer to the earth will be seen in this way to go at a slower rate than the other. Since the result had to hold for any types of accurate clocks, Einstein, in his imagination, used glowing atoms as the clocks, their rates being the frequencies of the light they emit. Thus atoms on the floor of the laboratory would be seen as giving off light of a lower frequency than that of the light given off by similar atoms on the ceiling.

The idea of using atoms as clocks allowed Einstein to apply the result to the sun. Since for the present purpose the gravitational effect of the earth is negligible compared with that of the much more massive sun, Einstein predicted that light from glowing atoms on the surface of the sun would be seen to have lower frequencies than light from similar atoms far from the sun. Since lowering the frequency shifts visible light toward the red end of the spectrum, this meant that the spectral lines of sunlight would be shifted toward the red compared with corresponding spectra generated in a laboratory on the earth. This shift of frequency is called the gravitational red shift. The amount of red shift—about one part in half a million—that Einstein calculated on the basis of his principle of equivalence turned out to be essentially the same as the amount he later obtained from his revolutionary theory of gravitation, which is called the general theory of relativity. Turbulence of the sun's surface made experimental verification difficult, but after a while the prediction was confirmed. And later, during the 1960s, the American experimenter R. V. Pound, with his students G. A. Rebka and J. L. Snider, actually managed to detect the much smaller gravitational red shift (about one part in 40 thousand million million) between the top and the bottom of a 22.5 meter tower at Harvard University.

The gravitational red shift does not arise from changes in the intrinsic rates of clocks. It arises from what befalls light signals as they traverse space and time in the presence of gravitation. Another hypothetical experiment proposed by Einstein told him more about the effect of gravitation on light. It is a simple experiment: Merely send a beam of light horizontally across the laboratory. Relative to the sky laboratory, because of that laboratory's upward acceleration, the ray will be curved downward. By the principle of equivalence, the corresponding ray in

the earth laboratory must be equally bent, and the cause there must be gravitation. Thus, gravitation bends light rays. For a ray of light grazing the sun, the amount of bending that Einstein calculated on the basis of his principle of equivalence turned out to be half the amount that he later obtained from his general theory of relativity. The verification of the bending will be discussed later.

There was a striking consequence of the gravitational bending of light rays. The waves of light corresponding to the rays would have to veer, and this could only happen if their speed of propagation higher up was greater than that lower down.

The discovery that gravitation affects the speed of light must have shaken Einstein. But he saw a way to turn it to possible advantage. With typical economy of means he decided to let the speed of light represent gravitation. The speed of light would then play the role of what is called the gravitational potential—a single number at each point in terms of which, in Newton's theory, one could specify the intensities of the gravitational forces everywhere. The idea did not work out. But since it took Einstein beyond the special theory of relativity, it was a valuable exploration, and it must have prepared him for the arduous years ahead.

We must pause here to note certain developments that had taken place in the field of geometry. Around the year 300 b.c., in his famous treatise on geometry called the *Elements,* Euclid built on a number of axioms and postulates that, for the most part, could hardly be doubted. His fifth postulate, however, seemed far less obvious than the others. In its original form it was a little subtle, but it is equivalent to saying the following: Consider a straight line and a point not on it, then in the plane containing the point and the line, there is only one straight line through the point that is parallel to the line.

———————————————•———————————————

————————————————————————————

Euclid's Parallel Postulate. We see only a part of the geometrically infinite entities represented by the light and heavy lines shown here. In the plane of the point and the given line, need there be precisely one light line parallel to the heavy line and passing through the heavy point as implied by Euclid's Postulate and as suggested by this figure or might there be many or perhaps none? The latter would be the case if every line in this plane that passes through the heavy point eventually crosses the heavy line, but possibly far outside the region shown here.

Through the ages, various mathematicians tried to prove this rather than take it on trust, but their proofs all turned out to be defective. Then, around 1823, two people hit on the same discovery. One of them was the Russian mathematician Nicolai Lobatschewsky, and the other the Hungarian mathematician Janos Bolyai. Their momentous discovery was that it is possible to have a self-consistent geometry in which, in the plane, there are not one but infinitely many lines through the point that are parallel to the given line. Actually, the great German mathematician Carl Friedrich Gauss had developed the same idea some years earlier but had withheld it from publication, apparently fearing that it would cause controversy.

The idea opened up new vistas. By exhibiting a specific self-consistent non-Euclidean geometry, the above mathematicians showed that the geometry of Euclid was not sacrosanct. The German mathematician Bernhard Riemann developed a different type of non-Euclidean geometry in which there were no parallel lines at all. These two types of non-Euclidean geometries, as usually expressed in two-dimensional form, could be envisaged as geometries pertaining to curved surfaces. For example, the non-Euclidean geometry proposed by Riemann could be thought of as the geometry belonging to the surface of a sphere where, playing the part of straight lines, were the great circles—the circles obtained by slicing the sphere with planes passing through its center. Arcs of great circles are the shortest surface distances and at the same time the straightest surface paths between points. Since every great circle intersects every other great circle, we see that there can be no parallel lines in this geometry.

We learn in school that the sum of the angles of a triangle is two right angles. What is not usually stressed is that in proving this theorem we have to use Euclid's parallel postulate in one or another of its forms. Indeed, so close is the link between the theorem and the postulate that we could regard the theorem itself as a postulate and then deduce the

Euclidean parallel postulate as a theorem. We should not be surprised, therefore, that in non-Euclidean geometry the theorem about the sum of the angles of a triangle does not hold.

Take for example, the non-Euclidean geometry proposed by Riemann pertaining to the surface of a sphere. For convenience, imagine lines of longitude and latitude drawn on it as if it were a globe representing the earth. The lines of longitude, being great circles, correspond to the straight lines of Euclidean geometry. But with one exception the lines of latitude are not great circles, the exception being the equator. A triangular figure on the surface of a sphere having sides that are segments of great circles of the sphere is called a spherical triangle.

Let us consider a special case. Take two different points, P and Q, on the equator. If, starting at P, we go due north, we travel along the line of longitude PN. If, starting at Q, we again go due north, we travel along the line of longitude QN. Since PN and QN both make right angles with the equator and are both aligned due north, we might be tempted to regard them as parallel despite the fact that there are no parallel lines in this type of non-Euclidean geometry. But PN and QN cannot be parallel, since they meet at N. Since the equator and the lines of longitude are great circles, the figure PQN is a spherical triangle. We have already remarked that the angles at P and Q are both right angles. It follows that the sum of the three angles of the spherical triangle PQN is greater than the two right angles, the excess in this case being the angle at N.

If we increase the angle at N, we automatically also increase the area of the spherical triangle. There happens to be a simple theorem that

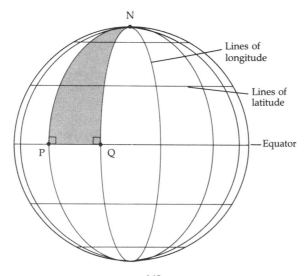

links the angles and surface area of a spherical triangle: in appropriate units, take the excess of the sum of the angles over two right angles and multiply that excess by the square of the radius of the sphere; then, according to the theorem, the result will be equal to the surface area of the spherical triangle.

So by drawing a spherical triangle on the surface of a sphere and measuring its angles and its surface area, we obtain data from which we can calculate the radius of the sphere. The important thing to note here is that the measurements of angles and area are intrinsic measurements—that is, measurements performed without leaving the two-dimensional surface of the sphere. We thus have the striking result that the radius of a sphere, and with it the degree of curvature of its surface, are intrinsic properties of that two-dimensional surface: they can be defined and dealt with solely in terms of the two-dimensional surface. There is another point worth mentioning here: The mere fact of the existence of a spherical triangle with two of its angles right angles is enough to indicate to us that the Pythagorean theorem cannot hold in the intrinsic non-Euclidean geometry belonging to the surface of a sphere. Actually, the theorem holds only in Euclidean geometry. But in various other geometries, it is approximately valid for small right triangles; the smaller the triangle, the closer the theorem to validity. In this sense we can say that the theorem is locally valid.

Gauss was well aware of the intrinsic nature of the curvature of the surface of a sphere. He showed that crucial aspects of the curvature of quite general surfaces with curvature varying from place to place were also intrinsic.

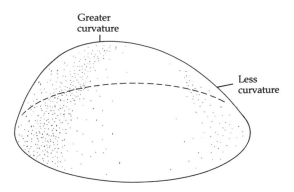

Greater
curvature

Less
curvature

Inspired by the intrinsic nature of the curvature of a surface, Riemann proposed an intrinsically curved three-dimensional space needing no additional dimension in which to curve or to be thought of as curving. Its curvature, which could vary from point to point, was implicit in measurements of lengths made wholly within it. From the curved two-dimensional and three-dimensional spaces, Riemann went on to envisage analogous intrinsically curved spaces having greater numbers of dimensions—four-dimensional spaces, five-dimensional spaces, and so on. These spaces proposed by Riemann are now called Riemannian spaces, and the geometry that pertains to such spaces is called Riemannian geometry. This Riemannian geometry, with its varying curvature, should not be confused with the simpler non-Euclidean geometry that Riemann proposed earlier.

Let us now return to Einstein. There is no need to trace out in detail here the advances and setbacks along the way to his general theory of relativity. Spurred by his general principle of relativity, he introduced what he called the principle of general covariance. It required that the laws of physics be expressed in a form that was the same for all systems of coordinates in space-time, and this posed a formidable mathematical problem that might well have been beyond Einstein's powers to solve— for he was a physicist, not a mathematician. But the mathematicians, with their extraordinary prescience, had already solved the underlying problem by developing the appropriate mathematical tool. And, as if it had all been fated, Marcell Grossmann, Einstein's fellow student at the Zurich Polytechnic Institute, happened to be an expert in non-Euclidean geometry and other relevant mathematical matters. Grossmann had become a professor at the Zurich Polytechnic Institute, and it was largely as a result of his urging that Einstein was brought back to the Institute in 1912. The two friends worked together, Grossmann being concerned with the mathematical aspects of their research and Einstein with the physical insights.

The mathematical tool that Einstein found waiting for him was called a tensor. Once we are familiar with the concept of a vector, which is a simple type of tensor, the basic idea of a tensor will pose no problem. Let us explore vectors for the case of two Euclidean dimensions and coordinate axes like those used on regular graph paper. On such graph paper the coordinates of a point give the distances from the coordinate axes. For example, the point with coordinates (1, 2) is 1 centimeter from the y axis and 2 centimeters from the x axis.

A vector has direction as well as magnitude, and we can represent it by drawing an arrow. Consider the particular vector that represents the displacement PQ from the point P to the point Q, where the coordinates of P are (1, 2) and those of Q are (4, 6). In this especially simple case, the differences $4 - 1$ and $6 - 2$ are called the components of the vector PQ. They turn out here to be 3 and 4, respectively, and they represent actual lengths. If we rotate the axes to a new orientation, the coordinates of P and Q will change, as will the differences that give the components of the vector. But although the components change, they still refer to the same vector: the vector does not change; the arrow does not change; only the components change. The same holds for more general cir-

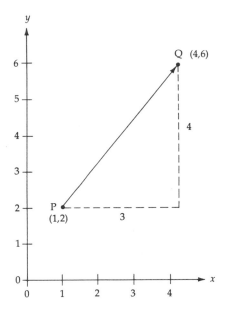

cumstances. A vector represents something that is independent of any coordinates we may introduce. It can thus represent something that is physically, objectively *there*.

In two dimensions a vector has two components in any given coordinate system, in three dimensions, three components in any given coordinate system, in four dimensions, four components, and so on. A general tensor differs from a vector basically in the number of components it has in a coordinate system, and also, alas, in the fact that it usually cannot be represented graphically by something analogous to the vector's arrow. But although the components of a tensor change when the coordinates are changed, the tensor itself does not change. Like the vector, it can thus represent something that is physically, objectively *there*. In four-dimensional space-time it turns out that the electric and magnetic aspects of the electromagnetic field combine to form a single tensor.

There is an especially important tensor called the metric tensor. To appreciate its geometric role, let us return to the case of two dimensions. Suppose that instead of regular graph paper we used a mesh of curved lines with numerical labels, as on page 146. The point denoted by P has the coordinates (1, 2), but these numbers are no longer actual distances. If we are dealing with geometry in a plane, such coordinate meshes are avoidable: we can always go back to regular graph paper and thus have coordinates that directly represent actual lengths. But what if

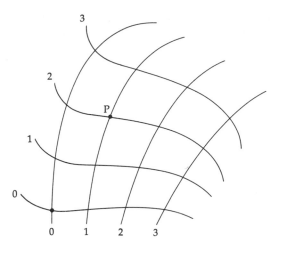

we were dealing with the intrinsic geometry on a curved surface on which regular graph paper would not fit smoothly? Then we could not use a coordinate mesh in which the coordinates all have direct metrical significance as distances. It is here that the metrical role of the metric tensor can be appreciated: for small differences of coordinates it converts the differences into actual lengths.

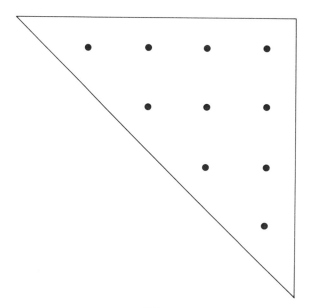

In two dimensions the metric tensor has $1 + 2 = 3$ independent components in any coordinate system. In three dimensions it has $1 + 2 + 3 = 6$ independent components. And in four-dimensional space-time it has $1 + 2 + 3 + 4 = 10$ independent components. It turns out that when one displays the ten independent components of the four-dimensional metric tensor, their locations, if represented by dots, automatically form a triangle pattern that is not unfamiliar to us.

Box 6.2

The metric tensor is thought of as a single entity but, as will be seen, it comes to us as collections of mathematical quantities called its components. Although we shall not be using the calculus here, it will be helpful in explaining the metric tensor if we use calculus symbols like dx and dy. We may think of the d in symbols of this sort as meaning a small change in x, or y, or whatever other quantity follows the d.

Consider, in two dimensions, two nearby points (x, y) and $(x + dx, y + dy)$. By the Pythagorean theorem, the small distance ds between them is given by

$$(ds)^2 = (dx)^2 + (dy)^2.$$

Mathematicians usually omit the parentheses and write this in the form

$$ds^2 = dx^2 + dy^2. \tag{1}$$

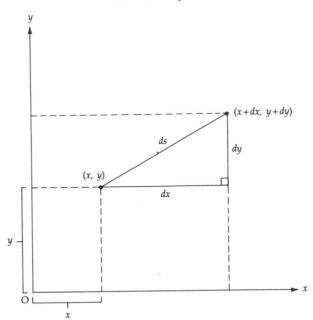

It may seem pointless to write this simple equation in the more complicated equivalent form

$$ds^2 = 1\,dx^2 + 0\,dx\,dy + 0\,dy\,dx + 1\,dy^2, \qquad (2)$$

but the 1s and 0s in this equation, which are invisible in equation (1), happen to be components of what is called the metric tensor. To see the general pattern of the components of this tensor clearly, let us borrow part of the standard tensor notation. Using suffixes 1 and 2 as labels, we rewrite equation (2), and thus also equation (1), in the form

$$ds^2 = g_{11}\,dx^2 + g_{12}\,dx\,dy + g_{21}\,dy\,dx + g_{22}\,dy^2, \qquad (3)$$

with g_{11} and g_{22} in this case (but not always) each standing for 1 and g_{12} and g_{21} for 0.

These four gs, collectively denoted by the symbol $g_{\mu\nu}$ and usually displayed in the square array

$$g_{11} \quad g_{12},$$
$$g_{21} \quad g_{22}$$

are the components of the metric tensor in the given coordinate system. Let us look at its role.

Suppose we change the scales along the x axis and the y axis by having, say, twice as many coordinate lines as before crossing the x axis and three times as many as before crossing the y axis. Then the old coordinates (x, y) of a point will be related to the new coordinates (x', y') of the point by $x = (\frac{1}{2})x'$, $y = (\frac{1}{3})y'$. So, instead of equation (1) we shall now have

$$ds^2 = ((\tfrac{1}{2})dx')^2 + ((\tfrac{1}{3})dy')^2 \qquad (4)$$
$$= (\tfrac{1}{4})dx'^2 + (\tfrac{1}{9})dy'^2,$$

which no longer has the simple Pythagorean form of equation (1). It turns out that under the present transformation of coordinates the components of the metric tensor change, its new components being just

$$g'_{11} = \tfrac{1}{4},\ g'_{22} = \tfrac{1}{9},\ g'_{12} = g'_{21} = 0.$$

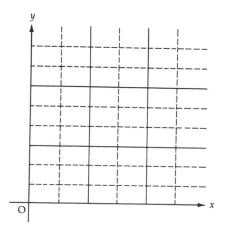

We see that equation (3) in primed coordinates amounts to equation (4). But whereas equation (4) does not have the same form as equation (1), equation (3), respectively with and without primes but otherwise unaltered in form, applies equally to equations (4) and (1) and represents essentially the same underlying situation.

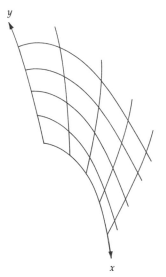

If we changed the graph paper so that the y axis was no longer perpendicular to the x axis, equation (1) would have to be replaced by an equation more complicated than equation (4), with the product $dx\,dy$ entering. But equation (3) would still retain its form, the new value of g_{12} (which has the same value as g_{21}) being now different from zero.

One further step. Suppose we went over to wavy, irregularly spaced coordinate lines. Then we would need a different substitute for (1) for each different location. But even in this case equation (3) would still retain its basic shape, the variability with respect to location being accounted for by the components of the metric tensor no longer being constants but instead being functions of the coordinates, that is, having, in general, different values in different places.

When we use coordinate systems more complicated than that used in (1), especially irregularly curved ones, the coordinates of points, and thus also the differences dx and dy of the coordinates of nearby points, are apt not to be directly related to distances. But in all coordinate systems the metric tensor lets us convert coordinate differences into actual distances—hence the name "metric". In all two-dimensional cases it does so by means of equation (3). And it does so even if we are dealing not with a plane but with the intrinsic geometry of a curved surface, even though the Pythagorean theorem is only valid within very small regions on such a surface. Moreover, the metric tensor contains all the intrinsic data needed to calculate at each point the intrinsic curvature of that surface.

In four dimensional space-time there are four coordinates instead of two, and the formula corresponding to (3) has sixteen terms on the right-hand side. The metric tensor, $g_{\mu\nu}$, has sixteen components:

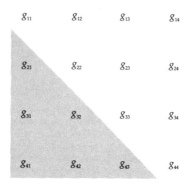

But since, for example, g_{31} is equal to g_{13}, the six components in the shaded triangle merely repeat corresponding components in the larger, unshaded triangle. Essentially, therefore, there are only ten independent components of the metric tensor of space-time. They tell us its intrinsic geometry, and by their means we can calculate the intrinsic curvature properties of space-time without the need for any extra dimensions.

It is a pity that the Pythagoreans did not know of the four-dimensional metric tensor. We recall their mystical veneration of the number ten and the triangle of dots representing it. They would surely have seen the arrangement of the metric tensor's components as a tilted image of their revered symbol. Imagine their sudden awe and sense of cosmic destiny at the moment of recognition.

If the Pythagoreans had been privileged to see and understand what Einstein created out of the metric tensor, they would have been enchanted by the beauty of his theory even though it conflicted with the general validity of the Pythagorean theorem. Perhaps, though, they would have rejoiced to see how Einstein's whole theory was, in a significant sense, an outgrowth of the metrical ideas embodied in that theorem.

Let us now look at the use Einstein made of the metric tensor of space-time. According to Newton's first law, a free particle in motion will continue to move with constant speed in a straight line. As has already been remarked, its world line in Minkowski space-time will thus also be straight. But such a line can be treated as analogous to the "shortest distance" between two points. It will thus involve the metric tensor. Einstein asked what this free motion would look like mathematically relative to the sky laboratory, and he noted that in terms of coordinates appropriate for the sky laboratory, the acceleration of the sky laboratory would change the components of the metric tensor. By the principle of equivalence the same must hold for the earth laboratory. And from this Einstein drew the momentous conclusion that, in his general theory of relativity, gravitation would have to be represented by the metric tensor of space-time.

A single gravitational potential had sufficed for Newtonian gravitation. But now, in a great leap forward, Einstein was proposing the use of ten gravitational potentials. This may seem an extravagance sadly lacking the cosmic simplicity that Einstein customarily sought in the universe. But in a significant sense Einstein was going not from one to ten but from one to none. For the metric tensor was already there in the underlying geometrical structure, and Einstein instead of introducing something extra was assigning to the metric tensor a dual role, one geometrical and the other gravitational. And this meant that gravitation was going to turn out to be not a force but something geometrical. It happens that the metric tensor contains all the data needed to calculate the intrinsic curvature of space-time. Since Einstein was representing gravitation by means of the metric tensor, it is not surprising that in his theory gravitation should turn out to be a curvature of space-time.

Like Maxwell's theory of electromagnetism, the new theory of gravitation was to be a field theory, with no instantaneous action at a distance. So, in place of Newton's inverse-square law of gravitation, Einstein and Grossmann looked for ten tensorial field equations that would in principle give the ten gravitational potentials—and thus also the space-time curvature—corresponding to any configuration of sun and planets or other sources of gravitation. To each configuration there should correspond its own unique gravitational field. But Einstein found a general proof that if one used tensor field equations for the metric tensor, it would follow that, corresponding to any given distribution of gravitating matter, there would inevitably be more than one gravitational field. It was a bitter blow. Einstein and his friend Grossmann struggled on, using equations that were somewhat but not wholly tensorial. They published a detailed paper—a sort of progress report—in 1913, and another in 1914, but after Einstein's departure for Berlin, the collaboration came to an end.

In Berlin Einstein labored on alone. The summer of 1914 saw the outbreak of World War I but Einstein, who was a Swiss citizen, continued working on his general theory of relativity in Berlin. And then, in 1915, came a sudden realization: There was an error in his proof against tensors.

At once Einstein returned to the idea of using tensor field equations, an idea that, as he said later, he had given up "only with a heavy heart." Now, suddenly, his theory crystallized into a thing of transcendent beauty. First let us look at the field equations, ten in number. Guided in part by analogy with an equation satisfied by the single gravitational potential in Newton's theory, Einstein sought equations expressible in terms of the metric tensor and its rates of change with respect to space and time. In the regular calculus, the rate of change of one quantity relative to another is found by a process called differentiation. It yields what is called a derivative. In the tensor calculus, the derivatives of tensors are, in general, not tensors. They therefore do not represent anything objective. Actually they represent a combination of the true rates of change and spurious rates of change that reflect nonuniformities of the coordinate mesh. The mathematicians who developed the tensor calculus found a way of filtering out the spurious rates of change by means of the metric tensor and thus obtaining the true rates of change of tensors. These objective rates are now called the "covariant derivatives" of tensors, but their earlier name is more vivid: "absolute derivatives." By either name they represent the uncontaminated rates of change of tensors, and it will therefore come as no surprise that the covariant derivatives of tensors are themselves tensors.

Thus in forming the field equations for gravitation, covariant derivatives of the metric tensor are the obvious choice. But its covariant derivatives turn out to be automatically zero. At first this seems to make it impossible to set up tensor field equations for the metric tensor. But the seeming calamity proves a blessing in disguise. Long before the advent of the very idea of a tensor, Gauss, in his studies of the intrinsic curvature of surfaces, found a mathematical expression that gave the curvature of a surface at any point. With the advent of the tensor calculus, the mathematical expression found by Gauss was seen to be a tensor made wholly out of the metric tensor and its ordinary derivatives. Riemann, and independently Elwin Christoffel, a professor at the Zurich Polytechnic Institute, had found the corresponding expression for curved spaces of any number of dimensions. It is now known as the curvature tensor.

In a significant sense the curvature tensor is unique. Precisely because the *covariant* derivatives of the metric tensor are all zero, if one seeks a tensor made wholly out of the metric tensor and its *regular* derivatives, one cannot avoid using the curvature tensor. In four dimensions the curvature tensor has twenty independent components. But from it, in a unique way, one can create a tensor having ten components, and if one seeks simplicity and adds a combined mathematical and physical condition to ensure both the possibility of solutions to the equations and the conservation of energy and momentum, then the ten field equations are uniquely determined. Thus, though he could not know it at the time, once Einstein had decided to use tensor equations and to represent gravitation solely by the metric tensor, the die was cast and the gravitational field equations followed essentially inevitably and without arbitrariness. Not that Einstein's path to the inevitable field equations was a direct one. He made errors and followed false leads—for example, he and Grossmann had considered but rejected a special case of the ultimate field equations—but always his intuition and sense of beauty brought him back to the right path.

The general theory of relativity differs markedly from the special theory of relativity. In the special theory there is no gravitation, and space-time is flat. In the general theory gravitation is present and space-time is curved. The curvature of space-time represents the gravitational field, the gravitational motions of bodies resulting from an intermeshing of their space-time curvatures.

For simple cases like the motion of a cannonball on the earth or the motion of a planet around the sun, it is convenient to treat the smaller object as what is called a test body—a body that reacts to a field but whose own contribution to the field is so small that it can safely be neglected. In the special theory of relativity, Newton's first law could be summed up by saying that the world line of a particle acted on by no forces was straight. In the general theory of relativity, because of the curvature of space-time, there are no straight world lines. However, some world lines are straighter than others. The straightest world lines, which are analogous to shortest distances, are called geodesics. What if one tried to come as close as possible to Newton's first law by saying that the world lines of test bodies acted on by no forces are geodesics? Then one would find that the curvature of space-time made the idea of gravitational force superfluous. The geodesics would give the motions of cannonballs, planets, and other such objects, with the curvature of space-time taking over the role formerly played by Newton's gravitational force. What of the gravitational bending of light rays? That, too, was derivable from the transplanted Newtonian law. Newton would have been delighted to see this beautiful extension of the domain of his first law of motion.

Like Newton's theory, the general theory of relativity had been given a dual structure. Corresponding to Newton's inverse-square law of gravitation were the field equations giving the gravitational field when the sources were specified; and corresponding to Newton's laws of motion was the geodesic hypothesis giving the motions—albeit of test bodies—when the gravitational field was specified. Much later, however, it was found that the geodesic hypothesis was not needed. The motions of bodies—and not just of test bodies—were already implicit in the field equations. Thus duality was replaced by unity, and with this unity came an even deeper beauty.

Once Einstein had his tensorial field equations—not yet in their final form, but in a form valid for his purposes—he calculated from them the gravitational curvature associated with the sun, and then, by means of the geodesic equations, he showed that while most of the planets would move in close accord with Newtonian predictions, there would be a particularly detectable deviation in the case of Mercury. Scientists had long known that because of the attractions of other planets and for other reasons, the motion of Mercury would not be in an ellipse but rather an ellipse that slowly rotated. That effect is called the rotation of

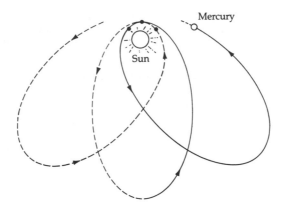

Advance of the perihelion of Mercury. The diagram is highly exaggerated.

the perihelion of Mercury. After all possible corrections had been applied in the Newtonian theory, there still remained an unexplained discrepancy: The ellipse was turning faster than it ought to by an amount of some 43 seconds of arc per century. Einstein found that his equations gave just this amount of deviation from the Newtonian prediction.

Box 6.3

> At each instant relative to the sun there is a three-dimensional space surrounding the sun. In order to indicate by a drawing the gravitational curvature of this space around the sun, we drop one spatial dimension so that we may represent three-dimensional space by a two-dimensional surface. A third dimension is thus available for indicating the intrinsic curvature.

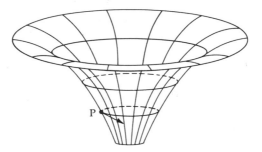

From the diagram we can see that if a planet wants to pursue a rectilinear path, the curvature will prevent it from doing so and will force it into an orbit around the sun.

The trouble is that we have no spare dimension available to represent the time dimension of space-time, and this makes the diagram somewhat unsatisfactory in spite of its vividness. The gravitational field of the sun, or of any other object, enters as an intrinsic curvature of four-dimensional space-time. The gravitation does not cause the curvature. It *is* the curvature.

The perihelion motion came out not only with the right numerical value but also in the correct direction. Furthermore, it emerged from the general theory of relativity in the most natural way, without forcing and without any special adjustments of numbers to make the result fit the observation. It was an outstanding triumph.

Einstein also calculated from his general theory of relativity the amount of the gravitational red shift between sun and earth and the amount by which light rays passing near the sun would be bent by the

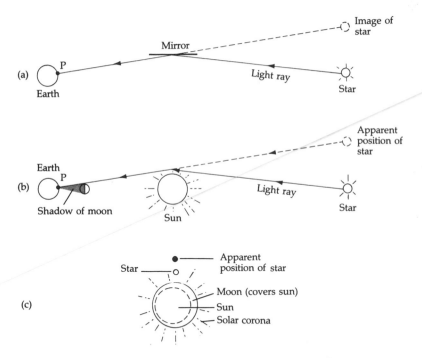

If we look at a star in a mirror as shown in (a), the image of the star will be displaced in the direction indicated by the dotted line. Analogously (b), the gravitational bending by the sun of a ray of light from a star causes the star to seem displaced away from the sun as indicated by the dotted line. An observer at point P in the shadow of the moon sees the sun completely covered by the moon, a total solar eclipse (c). With the direct sunlight blocked from his view, the sky becomes darker and he is able to see the stars, including those near the eclipse, where otherwise he could not. Thus the necessity at that time of observing gravitational bending during a solar eclipse. Nowadays, by means of quasars, one can dispense with the eclipse.

sun's gravitational curvature of space-time. As has already been mentioned, the full theory gave essentially the same red shift as that given by the principle of equivalence, while the gravitational bending of light rays came out as twice the earlier value. For a light ray grazing the sun, the general theory of relativity predicted a deflection of 1.7 seconds of arc—about ¼₀₀₀ of the angular width of the sun as seen from the earth.

With the war still raging, Einstein reported his researches to the Prussian Academy of Sciences, and they were published in its *Proceedings*. The details of Einstein's theory came, by way of neutral Holland, to the English scientist Arthur Eddington, who was so captivated by its beauty that, with the astronomer Royal, Frank Dyson, he began making plans for eclipse expeditions to test the predicted gravitational bending of light rays, his hope being to confirm the theory, not to disprove it. In 1919, at the time of the eclipse, fighting had ceased, but a state of war still existed, and hatred of the enemy was strong on both sides. One expedition went to Sobral in Brazil. The other, headed by Eddington, went to the West African island of Principe. Through telescopes, in the few minutes of total eclipse, photos were taken of stars in the angular neighborhood of the eclipsed sun. Eddington hastened to measure the displacements of the stars that would signal the bending of the light rays passing close to the sun on their journey from the stars to the earth. To his delight, his preliminary measurements supported Einstein's theory.

Later, back in England, the results of both expeditions were carefully studied, and the verdict again was favorable. Accordingly, the Royal Society of London—where Newton had presided two centuries before—invited the members of the Royal Astronomical Society to a joint meeting. There, with that gift for ceremony for which the British are famous, the favorable results were formally announced, and Einstein was hailed as a genius whose theory had successfully challenged that of the great Isaac Newton. The newspapers reported the historic event, and Einstein became world famous overnight.

The story of cosmological and other applications of Einstein's general theory of relativity belongs to a different book. So, too, does the story of the attempts to unify gravitation with electromagnetism and other fundamental forces of nature. I have been telling here of the development of the concept of relativity through the ages, and the present book is now nearing its end. In the years since the general theory was propounded, it has met all experimental challenges. It towers as a monumental masterwork—of art as well as science. Its artistry resides in its inevitability, the economy of its structure, the basic simplicity that shines through its complexities, and a pervasive beauty that, like all beauty, defies analysis.

What is it that pulls the apple to the ground, bends the circling moon to the earth, and makes the planets captives of the sun? It is no longer a force acting at a distance, but something far more primordial. It is intangible time and space themselves, acting in awesome concert as curved space-time and holding sway over all things in the universe.

When the eclipse observations in 1919 confirmed Einstein's prediction of the gravitational bending of light rays, he was naturally elated. But it is worthwhile to look at the matter more closely. In wartime Germany news of Einstein's theory had reached the public long before the results of the eclipse expeditions had been announced. In 1916, at the request of a German publisher, Einstein wrote for the layman an account of his special and general theories of relativity. At that time neither the red shift nor the bending of light rays had been experimentally verified. In his book, after telling of the successful calculation of the perihelion motion of Mercury, Einstein wrote, referring to the red shift and the bending of light rays: "I do not doubt that these deductions from the theory will be confirmed also."

We could infer from this remark that the perihelion calculation was the prime source of Einstein's confidence. But this would underrate Einstein. The main source of his confidence was the profound simplicity and beauty of his theory. To see this, let us look at the pace and manner of its sudden crystallization once he returned to the idea of using tensor equations.

On November 4, 1915, at one of the weekly meetings of the Prussian Academy of Sciences, Einstein presented a technical paper to his fellow professionals telling of his current work on the general theory of relativity. In it he had still not quite attained his goal of using tensor equations. The next week, November 11, he presented a further paper, in which he dealt more specifically with tensors. On November 18 he presented his calculations accounting for the residual motion of the perihelion of Mercury. And the following week, on November 25, he added finishing touches to the field equations and presented the theory in full tensorial beauty: The crystallization was complete.

With this chronology in mind, let us return to his paper of November 4. At that stage Einstein knew that, if one ignored very small effects, his equations would give Newtonian results. He knew, too, that by building on his principle of equivalence he had incorporated a profoundly simple interpretation of Galileo's law of fall. But the eclipse verification of the bending of light rays was far in the future, as was the verification of the red shift. The important perihelion calculations had yet to be made, and the tensors were not yet fully incorporated. Nevertheless, in this highly mathematical report to fellow scientists we find revealing and extraordinary words that testify to the prime role of aesthetic considerations in Einstein's science. Here are those words:

"Hardly anyone who has truly understood this theory will be able to resist being captivated by its magic."

Only a man with powerful aesthetic intuition could have created the general theory of relativity. The work had not been easy. Here is what Einstein said in 1934 in an article telling of his path to the theory:

> In the light of knowledge attained, the happy achievement seems almost a matter of course, and any intelligent student can grasp it without too much trouble. But the years of anxious searching in the dark with their intense longing, their alternations of confidence and exhaustion and the final emergence into the light—only those who have experienced it can understand that.

Index

159

A CATALOG OF SELECTED
DOVER BOOKS
IN ALL FIELDS OF INTEREST

A CATALOG OF SELECTED DOVER
BOOKS IN ALL FIELDS OF INTEREST

CONCERNING THE SPIRITUAL IN ART, Wassily Kandinsky. Pioneering work by father of abstract art. Thoughts on color theory, nature of art. Analysis of earlier masters. 12 illustrations. 80pp. of text. 5⅜ x 8½. 23411-8 Pa. $3.95

ANIMALS: 1,419 Copyright-Free Illustrations of Mammals, Birds, Fish, Insects, etc., Jim Harter (ed.). Clear wood engravings present, in extremely lifelike poses, over 1,000 species of animals. One of the most extensive pictorial sourcebooks of its kind. Captions. Index. 284pp. 9 x 12. 23766-4 Pa. $12.95

CELTIC ART: The Methods of Construction, George Bain. Simple geometric techniques for making Celtic interlacements, spirals, Kells-type initials, animals, humans, etc. Over 500 illustrations. 160pp. 9 x 12. (USO) 22923-8 Pa. $9.95

AN ATLAS OF ANATOMY FOR ARTISTS, Fritz Schider. Most thorough reference work on art anatomy in the world. Hundreds of illustrations, including selections from works by Vesalius, Leonardo, Goya, Ingres, Michelangelo, others. 593 illustrations. 192pp. 7⅛ x 10¼. 20241-0 Pa. $9.95

CELTIC HAND STROKE-BY-STROKE (Irish Half-Uncial from "The Book of Kells"): An Arthur Baker Calligraphy Manual, Arthur Baker. Complete guide to creating each letter of the alphabet in distinctive Celtic manner. Covers hand position, strokes, pens, inks, paper, more. Illustrated. 48pp. 8¼ x 11. 24336-2 Pa. $3.95

EASY ORIGAMI, John Montroll. Charming collection of 32 projects (hat, cup, pelican, piano, swan, many more) specially designed for the novice origami hobbyist. Clearly illustrated easy-to-follow instructions insure that even beginning papercrafters will achieve successful results. 48pp. 8¼ x 11. 27298-2 Pa. $3.50

THE COMPLETE BOOK OF BIRDHOUSE CONSTRUCTION FOR WOOD-WORKERS, Scott D. Campbell. Detailed instructions, illustrations, tables. Also data on bird habitat and instinct patterns. Bibliography. 3 tables. 63 illustrations in 15 figures. 48pp. 5¼ x 8½. 24407-5 Pa. $2.50

BLOOMINGDALE'S ILLUSTRATED 1886 CATALOG: Fashions, Dry Goods and Housewares, Bloomingdale Brothers. Famed merchants' extremely rare catalog depicting about 1,700 products: clothing, housewares, firearms, dry goods, jewelry, more. Invaluable for dating, identifying vintage items. Also, copyright-free graphics for artists, designers. Co-published with Henry Ford Museum & Greenfield Village. 160pp. 8¼ x 11. 25780-0 Pa. $10.95

HISTORIC COSTUME IN PICTURES, Braun & Schneider. Over 1,450 costumed figures in clearly detailed engravings—from dawn of civilization to end of 19th century. Captions. Many folk costumes. 256pp. 8⅜ x 11¾. 23150-X Pa. $12.95

STICKLEY CRAFTSMAN FURNITURE CATALOGS, Gustav Stickley and L. & J. G. Stickley. Beautiful, functional furniture in two authentic catalogs from 1910. 594 illustrations, including 277 photos, show settles, rockers, armchairs, reclining chairs, bookcases, desks, tables. 183pp. 6½ x 9¼. 23838-5 Pa. $9.95

AMERICAN LOCOMOTIVES IN HISTORIC PHOTOGRAPHS: 1858 to 1949, Ron Ziel (ed.). A rare collection of 126 meticulously detailed official photographs, called "builder portraits," of American locomotives that majestically chronicle the rise of steam locomotive power in America. Introduction. Detailed captions. xi + 129pp. 9 x 12. 27393-8 Pa. $12.95

AMERICA'S LIGHTHOUSES: An Illustrated History, Francis Ross Holland, Jr. Delightfully written, profusely illustrated fact-filled survey of over 200 American lighthouses since 1716. History, anecdotes, technological advances, more. 240pp. 8 x 10¾. 25576-X Pa. $12.95

TOWARDS A NEW ARCHITECTURE, Le Corbusier. Pioneering manifesto by founder of "International School." Technical and aesthetic theories, views of industry, economics, relation of form to function, "mass-production split" and much more. Profusely illustrated. 320pp. 6⅛ x 9¼. (USO) 25023-7 Pa. $9.95

HOW THE OTHER HALF LIVES, Jacob Riis. Famous journalistic record, exposing poverty and degradation of New York slums around 1900, by major social reformer. 100 striking and influential photographs. 233pp. 10 x 7⅞. 22012-5 Pa. $10.95

FRUIT KEY AND TWIG KEY TO TREES AND SHRUBS, William M. Harlow. One of the handiest and most widely used identification aids. Fruit key covers 120 deciduous and evergreen species; twig key 160 deciduous species. Easily used. Over 300 photographs. 126pp. 5⅜ x 8½. 20511-8 Pa. $3.95

COMMON BIRD SONGS, Dr. Donald J. Borror. Songs of 60 most common U.S. birds: robins, sparrows, cardinals, bluejays, finches, more—arranged in order of increasing complexity. Up to 9 variations of songs of each species. Cassette and manual 99911-4 $8.95

ORCHIDS AS HOUSE PLANTS, Rebecca Tyson Northen. Grow cattleyas and many other kinds of orchids—in a window, in a case, or under artificial light. 63 illustrations. 148pp. 5⅜ x 8½. 23261-1 Pa. $4.95

MONSTER MAZES, Dave Phillips. Masterful mazes at four levels of difficulty. Avoid deadly perils and evil creatures to find magical treasures. Solutions for all 32 exciting illustrated puzzles. 48pp. 8¼ x 11. 26005-4 Pa. $2.95

MOZART'S DON GIOVANNI (DOVER OPERA LIBRETTO SERIES), Wolfgang Amadeus Mozart. Introduced and translated by Ellen H. Bleiler. Standard Italian libretto, with complete English translation. Convenient and thoroughly portable—an ideal companion for reading along with a recording or the performance itself. Introduction. List of characters. Plot summary. 121pp. 5¼ x 8½. 24944-1 Pa. $2.95

TECHNICAL MANUAL AND DICTIONARY OF CLASSICAL BALLET, Gail Grant. Defines, explains, comments on steps, movements, poses and concepts. 15-page pictorial section. Basic book for student, viewer. 127pp. 5⅜ x 8½. 21843-0 Pa. $4.95

BRASS INSTRUMENTS: Their History and Development, Anthony Baines. Authoritative, updated survey of the evolution of trumpets, trombones, bugles, cornets, French horns, tubas and other brass wind instruments. Over 140 illustrations and 48 music examples. Corrected and updated by author. New preface. Bibliography. 320pp. 5⅜ x 8½. 27574-4 Pa. $9.95

HOLLYWOOD GLAMOR PORTRAITS, John Kobal (ed.). 145 photos from 1926-49. Harlow, Gable, Bogart, Bacall; 94 stars in all. Full background on photographers, technical aspects. 160pp. 8⅜ x 11¼. 23352-9 Pa. $12.95

MAX AND MORITZ, Wilhelm Busch. Great humor classic in both German and English. Also 10 other works: "Cat and Mouse," "Plisch and Plumm," etc. 216pp. 5⅜ x 8½. 20181-3 Pa. $6.95

THE RAVEN AND OTHER FAVORITE POEMS, Edgar Allan Poe. Over 40 of the author's most memorable poems: "The Bells," "Ulalume," "Israfel," "To Helen," "The Conqueror Worm," "Eldorado," "Annabel Lee," many more. Alphabetic lists of titles and first lines. 64pp. 5$\frac{3}{16}$ x 8¼. 26685-0 Pa. $1.00

PERSONAL MEMOIRS OF U. S. GRANT, Ulysses Simpson Grant. Intelligent, deeply moving firsthand account of Civil War campaigns, considered by many the finest military memoirs ever written. Includes letters, historic photographs, maps and more. 528pp. 6⅛ x 9¼. 28587-1 Pa. $11.95

AMULETS AND SUPERSTITIONS, E. A. Wallis Budge. Comprehensive discourse on origin, powers of amulets in many ancient cultures: Arab, Persian Babylonian, Assyrian, Egyptian, Gnostic, Hebrew, Phoenician, Syriac, etc. Covers cross, swastika, crucifix, seals, rings, stones, etc. 584pp. 5⅜ x 8½. 23573-4 Pa. $12.95

RUSSIAN STORIES/PYCCKNE PACCKA3bl: A Dual-Language Book, edited by Gleb Struve. Twelve tales by such masters as Chekhov, Tolstoy, Dostoevsky, Pushkin, others. Excellent word-for-word English translations on facing pages, plus teaching and study aids, Russian/English vocabulary, biographical/critical introductions, more. 416pp. 5⅜ x 8½. 26244-8 Pa. $8.95

PHILADELPHIA THEN AND NOW: 60 Sites Photographed in the Past and Present, Kenneth Finkel and Susan Oyama. Rare photographs of City Hall, Logan Square, Independence Hall, Betsy Ross House, other landmarks juxtaposed with contemporary views. Captures changing face of historic city. Introduction. Captions. 128pp. 8¼ x 11. 25790-8 Pa. $9.95

AIA ARCHITECTURAL GUIDE TO NASSAU AND SUFFOLK COUNTIES, LONG ISLAND, The American Institute of Architects, Long Island Chapter, and the Society for the Preservation of Long Island Antiquities. Comprehensive, well-researched and generously illustrated volume brings to life over three centuries of Long Island's great architectural heritage. More than 240 photographs with authoritative, extensively detailed captions. 176pp. 8¼ x 11. 26946-9 Pa. $14.95

NORTH AMERICAN INDIAN LIFE: Customs and Traditions of 23 Tribes, Elsie Clews Parsons (ed.). 27 fictionalized essays by noted anthropologists examine religion, customs, government, additional facets of life among the Winnebago, Crow, Zuni, Eskimo, other tribes. 480pp. 6⅛ x 9¼. 27377-6 Pa. $10.95

FRANK LLOYD WRIGHT'S HOLLYHOCK HOUSE, Donald Hoffmann. Lavishly illustrated, carefully documented study of one of Wright's most controversial residential designs. Over 120 photographs, floor plans, elevations, etc. Detailed perceptive text by noted Wright scholar. Index. 128pp. 9¼ x 10¾. 27133-1 Pa. $11.95

THE MALE AND FEMALE FIGURE IN MOTION: 60 Classic Photographic Sequences, Eadweard Muybridge. 60 true-action photographs of men and women walking, running, climbing, bending, turning, etc., reproduced from rare 19th-century masterpiece. vi + 121pp. 9 x 12. 24745-7 Pa. $10.95

1001 QUESTIONS ANSWERED ABOUT THE SEASHORE, N. J. Berrill and Jacquelyn Berrill. Queries answered about dolphins, sea snails, sponges, starfish, fishes, shore birds, many others. Covers appearance, breeding, growth, feeding, much more. 305pp. 5¼ x 8¼. 23366-9 Pa. $8.95

GUIDE TO OWL WATCHING IN NORTH AMERICA, Donald S. Heintzelman. Superb guide offers complete data and descriptions of 19 species: barn owl, screech owl, snowy owl, many more. Expert coverage of owl-watching equipment, conservation, migrations and invasions, etc. Guide to observing sites. 84 illustrations. xiii + 193pp. 5⅜ x 8½. 27344-X Pa. $8.95

MEDICINAL AND OTHER USES OF NORTH AMERICAN PLANTS: A Historical Survey with Special Reference to the Eastern Indian Tribes, Charlotte Erichsen-Brown. Chronological historical citations document 500 years of usage of plants, trees, shrubs native to eastern Canada, northeastern U.S. Also complete identifying information. 343 illustrations. 544pp. 6½ x 9¼. 25951-X Pa. $12.95

STORYBOOK MAZES, Dave Phillips. 23 stories and mazes on two-page spreads: Wizard of Oz, Treasure Island, Robin Hood, etc. Solutions. 64pp. 8¼ x 11. 23628-5 Pa. $2.95

NEGRO FOLK MUSIC, U.S.A., Harold Courlander. Noted folklorist's scholarly yet readable analysis of rich and varied musical tradition. Includes authentic versions of over 40 folk songs. Valuable bibliography and discography. xi + 324pp. 5⅜ x 8½. 27350-4 Pa. $9.95

MOVIE-STAR PORTRAITS OF THE FORTIES, John Kobal (ed.). 163 glamor, studio photos of 106 stars of the 1940s: Rita Hayworth, Ava Gardner, Marlon Brando, Clark Gable, many more. 176pp. 8⅜ x 11¼. 23546-7 Pa. $12.95

BENCHLEY LOST AND FOUND, Robert Benchley. Finest humor from early 30s, about pet peeves, child psychologists, post office and others. Mostly unavailable elsewhere. 73 illustrations by Peter Arno and others. 183pp. 5⅜ x 8½. 22410-4 Pa. $6.95

YEKL and THE IMPORTED BRIDEGROOM AND OTHER STORIES OF YIDDISH NEW YORK, Abraham Cahan. Film Hester Street based on Yekl (1896). Novel, other stories among first about Jewish immigrants on N.Y.'s East Side. 240pp. 5⅜ x 8½. 22427-9 Pa. $6.95

SELECTED POEMS, Walt Whitman. Generous sampling from *Leaves of Grass*. Twenty-four poems include "I Hear America Singing," "Song of the Open Road," "I Sing the Body Electric," "When Lilacs Last in the Dooryard Bloom'd," "O Captain! My Captain!"—all reprinted from an authoritative edition. Lists of titles and first lines. 128pp. 5³⁄₁₆ x 8¼. 26878-0 Pa. $1.00

THE BEST TALES OF HOFFMANN, E. T. A. Hoffmann. 10 of Hoffmann's most important stories: "Nutcracker and the King of Mice," "The Golden Flowerpot," etc. 458pp. 5⅜ x 8½. 21793-0 Pa. $9.95

FROM FETISH TO GOD IN ANCIENT EGYPT, E. A. Wallis Budge. Rich detailed survey of Egyptian conception of "God" and gods, magic, cult of animals, Osiris, more. Also, superb English translations of hymns and legends. 240 illustrations. 545pp. 5⅜ x 8½. 25803-3 Pa. $13.95

FRENCH STORIES/CONTES FRANÇAIS: A Dual-Language Book, Wallace Fowlie. Ten stories by French masters, Voltaire to Camus: "Micromegas" by Voltaire; "The Atheist's Mass" by Balzac; "Minuet" by de Maupassant; "The Guest" by Camus, six more. Excellent English translations on facing pages. Also French-English vocabulary list, exercises, more. 352pp. 5⅜ x 8½. 26443-2 Pa. $8.95

CHICAGO AT THE TURN OF THE CENTURY IN PHOTOGRAPHS: 122 Historic Views from the Collections of the Chicago Historical Society, Larry A. Viskochil. Rare large-format prints offer detailed views of City Hall, State Street, the Loop, Hull House, Union Station, many other landmarks, circa 1904-1913. Introduction. Captions. Maps. 144pp. 9⅜ x 12¼. 24656-6 Pa. $12.95

OLD BROOKLYN IN EARLY PHOTOGRAPHS, 1865-1929, William Lee Younger. Luna Park, Gravesend race track, construction of Grand Army Plaza, moving of Hotel Brighton, etc. 157 previously unpublished photographs. 165pp. 8⅞ x 11¾. 23587-4 Pa. $13.95

THE MYTHS OF THE NORTH AMERICAN INDIANS, Lewis Spence. Rich anthology of the myths and legends of the Algonquins, Iroquois, Pawnees and Sioux, prefaced by an extensive historical and ethnological commentary. 36 illustrations. 480pp. 5⅜ x 8½. 25967-6 Pa. $8.95

AN ENCYCLOPEDIA OF BATTLES: Accounts of Over 1,560 Battles from 1479 B.C. to the Present, David Eggenberger. Essential details of every major battle in recorded history from the first battle of Megiddo in 1479 B.C. to Grenada in 1984. List of Battle Maps. New Appendix covering the years 1967-1984. Index. 99 illustrations. 544pp. 6½ x 9¼. 24913-1 Pa. $14.95

SAILING ALONE AROUND THE WORLD, Captain Joshua Slocum. First man to sail around the world, alone, in small boat. One of great feats of seamanship told in delightful manner. 67 illustrations. 294pp. 5⅜ x 8½. 20326-3 Pa. $5.95

ANARCHISM AND OTHER ESSAYS, Emma Goldman. Powerful, penetrating, prophetic essays on direct action, role of minorities, prison reform, puritan hypocrisy, violence, etc. 271pp. 5⅜ x 8½. 22484-8 Pa. $6.95

MYTHS OF THE HINDUS AND BUDDHISTS, Ananda K. Coomaraswamy and Sister Nivedita. Great stories of the epics; deeds of Krishna, Shiva, taken from puranas, Vedas, folk tales; etc. 32 illustrations. 400pp. 5⅜ x 8½. 21759-0 Pa. $10.95

BEYOND PSYCHOLOGY, Otto Rank. Fear of death, desire of immortality, nature of sexuality, social organization, creativity, according to Rankian system. 291pp. 5⅜ x 8½. 20485-5 Pa. $8.95

A THEOLOGICO-POLITICAL TREATISE, Benedict Spinoza. Also contains unfinished Political Treatise. Great classic on religious liberty, theory of government on common consent. R. Elwes translation. Total of 421pp. 5⅜ x 8½. 20249-6 Pa. $9.95

MY BONDAGE AND MY FREEDOM, Frederick Douglass. Born a slave, Douglass became outspoken force in antislavery movement. The best of Douglass' autobiographies. Graphic description of slave life. 464pp. 5⅜ x 8½. 22457-0 Pa. $8.95

FOLLOWING THE EQUATOR: A Journey Around the World, Mark Twain. Fascinating humorous account of 1897 voyage to Hawaii, Australia, India, New Zealand, etc. Ironic, bemused reports on peoples, customs, climate, flora and fauna, politics, much more. 197 illustrations. 720pp. 5⅜ x 8½.   26113-1 Pa. $15.95

THE PEOPLE CALLED SHAKERS, Edward D. Andrews. Definitive study of Shakers: origins, beliefs, practices, dances, social organization, furniture and crafts, etc. 33 illustrations. 351pp. 5⅜ x 8½.   21081-2 Pa. $8.95

THE MYTHS OF GREECE AND ROME, H. A. Guerber. A classic of mythology, generously illustrated, long prized for its simple, graphic, accurate retelling of the principal myths of Greece and Rome, and for its commentary on their origins and significance. With 64 illustrations by Michelangelo, Raphael, Titian, Rubens, Canova, Bernini and others. 480pp. 5⅜ x 8½.   27584-1 Pa. $9.95

PSYCHOLOGY OF MUSIC, Carl E. Seashore. Classic work discusses music as a medium from psychological viewpoint. Clear treatment of physical acoustics, auditory apparatus, sound perception, development of musical skills, nature of musical feeling, host of other topics. 88 figures. 408pp. 5⅜ x 8½.   21851-1 Pa. $10.95

THE PHILOSOPHY OF HISTORY, Georg W. Hegel. Great classic of Western thought develops concept that history is not chance but rational process, the evolution of freedom. 457pp. 5⅜ x 8½.   20112-0 Pa. $9.95

THE BOOK OF TEA, Kakuzo Okakura. Minor classic of the Orient: entertaining, charming explanation, interpretation of traditional Japanese culture in terms of tea ceremony. 94pp. 5⅜ x 8½.   20070-1 Pa. $3.95

LIFE IN ANCIENT EGYPT, Adolf Erman. Fullest, most thorough, detailed older account with much not in more recent books, domestic life, religion, magic, medicine, commerce, much more. Many illustrations reproduce tomb paintings, carvings, hieroglyphs, etc. 597pp. 5⅜ x 8½.   22632-8 Pa. $11.95

SUNDIALS, Their Theory and Construction, Albert Waugh. Far and away the best, most thorough coverage of ideas, mathematics concerned, types, construction, adjusting anywhere. Simple, nontechnical treatment allows even children to build several of these dials. Over 100 illustrations. 230pp. 5⅜ x 8½.   22947-5 Pa. $7.95

DYNAMICS OF FLUIDS IN POROUS MEDIA, Jacob Bear. For advanced students of ground water hydrology, soil mechanics and physics, drainage and irrigation engineering, and more. 335 illustrations. Exercises, with answers. 784pp. 6⅛ x 9¼.
                                                       65675-6 Pa. $19.95

SONGS OF EXPERIENCE: Facsimile Reproduction with 26 Plates in Full Color, William Blake. 26 full-color plates from a rare 1826 edition. Includes "The Tyger," "London," "Holy Thursday," and other poems. Printed text of poems. 48pp. 5¼ x 7.
                                                       24636-1 Pa. $4.95

OLD-TIME VIGNETTES IN FULL COLOR, Carol Belanger Grafton (ed.). Over 390 charming, often sentimental illustrations, selected from archives of Victorian graphics–pretty women posing, children playing, food, flowers, kittens and puppies, smiling cherubs, birds and butterflies, much more. All copyright-free. 48pp. 9¼ x 12¼.
                                                       27269-9 Pa. $7.95

PERSPECTIVE FOR ARTISTS, Rex Vicat Cole. Depth, perspective of sky and sea, shadows, much more, not usually covered. 391 diagrams, 81 reproductions of drawings and paintings. 279pp. 5⅜ x 8½. 22487-2 Pa. $7.95

DRAWING THE LIVING FIGURE, Joseph Sheppard. Innovative approach to artistic anatomy focuses on specifics of surface anatomy, rather than muscles and bones. Over 170 drawings of live models in front, back and side views, and in widely varying poses. Accompanying diagrams. 177 illustrations. Introduction. Index. 144pp. 8⅜ x11¼. 26723-7 Pa. $8.95

GOTHIC AND OLD ENGLISH ALPHABETS: 100 Complete Fonts, Dan X. Solo. Add power, elegance to posters, signs, other graphics with 100 stunning copyright-free alphabets: Blackstone, Dolbey, Germania, 97 more—including many lower-case, numerals, punctuation marks. 104pp. 8⅛ x 11. 24695-7 Pa. $8.95

HOW TO DO BEADWORK, Mary White. Fundamental book on craft from simple projects to five-bead chains and woven works. 106 illustrations. 142pp. 5⅜ x 8. 20697-1 Pa. $4.95

THE BOOK OF WOOD CARVING, Charles Marshall Sayers. Finest book for beginners discusses fundamentals and offers 34 designs. "Absolutely first rate . . . well thought out and well executed."–E. J. Tangerman. 118pp. 7¾ x 10⅝. 23654-4 Pa. $6.95

ILLUSTRATED CATALOG OF CIVIL WAR MILITARY GOODS: Union Army Weapons, Insignia, Uniform Accessories, and Other Equipment, Schuyler, Hartley, and Graham. Rare, profusely illustrated 1846 catalog includes Union Army uniform and dress regulations, arms and ammunition, coats, insignia, flags, swords, rifles, etc. 226 illustrations. 160pp. 9 x 12. 24939-5 Pa. $10.95

WOMEN'S FASHIONS OF THE EARLY 1900s: An Unabridged Republication of "New York Fashions, 1909," National Cloak & Suit Co. Rare catalog of mail-order fashions documents women's and children's clothing styles shortly after the turn of the century. Captions offer full descriptions, prices. Invaluable resource for fashion, costume historians. Approximately 725 illustrations. 128pp. 8⅜ x 11¼. 27276-1 Pa. $11.95

THE 1912 AND 1915 GUSTAV STICKLEY FURNITURE CATALOGS, Gustav Stickley. With over 200 detailed illustrations and descriptions, these two catalogs are essential reading and reference materials and identification guides for Stickley furniture. Captions cite materials, dimensions and prices. 112pp. 6½ x 9¼. 26676-1 Pa. $9.95

EARLY AMERICAN LOCOMOTIVES, John H. White, Jr. Finest locomotive engravings from early 19th century: historical (1804–74), main-line (after 1870), special, foreign, etc. 147 plates. 142pp. 11⅜ x 8¼. 22772-3 Pa. $10.95

THE TALL SHIPS OF TODAY IN PHOTOGRAPHS, Frank O. Braynard. Lavishly illustrated tribute to nearly 100 majestic contemporary sailing vessels: Amerigo Vespucci, Clearwater, Constitution, Eagle, Mayflower, Sea Cloud, Victory, many more. Authoritative captions provide statistics, background on each ship. 190 black-and-white photographs and illustrations. Introduction. 128pp. 8⅞ x 11¾. 27163-3 Pa. $13.95

EARLY NINETEENTH-CENTURY CRAFTS AND TRADES, Peter Stockham (ed.). Extremely rare 1807 volume describes to youngsters the crafts and trades of the day: brickmaker, weaver, dressmaker, bookbinder, ropemaker, saddler, many more. Quaint prose, charming illustrations for each craft. 20 black-and-white line illustrations. 192pp. 4⅝ x 6. 27293-1 Pa. $4.95

VICTORIAN FASHIONS AND COSTUMES FROM HARPER'S BAZAR, 1867–1898, Stella Blum (ed.). Day costumes, evening wear, sports clothes, shoes, hats, other accessories in over 1,000 detailed engravings. 320pp. 9⅜ x 12¼. 22990-4 Pa. $14.95

GUSTAV STICKLEY, THE CRAFTSMAN, Mary Ann Smith. Superb study surveys broad scope of Stickley's achievement, especially in architecture. Design philosophy, rise and fall of the Craftsman empire, descriptions and floor plans for many Craftsman houses, more. 86 black-and-white halftones. 31 line illustrations. Introduction 208pp. 6½ x 9¼. 27210-9 Pa. $9.95

THE LONG ISLAND RAIL ROAD IN EARLY PHOTOGRAPHS, Ron Ziel. Over 220 rare photos, informative text document origin (1844) and development of rail service on Long Island. Vintage views of early trains, locomotives, stations, passengers, crews, much more. Captions. 8⅞ x 11¾. 26301-0 Pa. $13.95

THE BOOK OF OLD SHIPS: From Egyptian Galleys to Clipper Ships, Henry B. Culver. Superb, authoritative history of sailing vessels, with 80 magnificent line illustrations. Galley, bark, caravel, longship, whaler, many more. Detailed, informative text on each vessel by noted naval historian. Introduction. 256pp. 5⅜ x 8½. 27332-6 Pa. $7.95

TEN BOOKS ON ARCHITECTURE, Vitruvius. The most important book ever written on architecture. Early Roman aesthetics, technology, classical orders, site selection, all other aspects. Morgan translation. 331pp. 5⅜ x 8½. 20645-9 Pa. $8.95

THE HUMAN FIGURE IN MOTION, Eadweard Muybridge. More than 4,500 stopped-action photos, in action series, showing undraped men, women, children jumping, lying down, throwing, sitting, wrestling, carrying, etc. 390pp. 7⅞ x 10⅝. 20204-6 Clothbd. $25.95

TREES OF THE EASTERN AND CENTRAL UNITED STATES AND CANADA, William M. Harlow. Best one-volume guide to 140 trees. Full descriptions, woodlore, range, etc. Over 600 illustrations. Handy size. 288pp. 4½ x 6⅜. 20395-6 Pa. $6.95

SONGS OF WESTERN BIRDS, Dr. Donald J. Borror. Complete song and call repertoire of 60 western species, including flycatchers, juncoes, cactus wrens, many more—includes fully illustrated booklet. Cassette and manual 99913-0 $8.95

GROWING AND USING HERBS AND SPICES, Milo Miloradovich. Versatile handbook provides all the information needed for cultivation and use of all the herbs and spices available in North America. 4 illustrations. Index. Glossary. 236pp. 5⅜ x 8½. 25058-X Pa. $6.95

BIG BOOK OF MAZES AND LABYRINTHS, Walter Shepherd. 50 mazes and labyrinths in all—classical, solid, ripple, and more—in one great volume. Perfect inexpensive puzzler for clever youngsters. Full solutions. 112pp. 8⅛ x 11. 22951-3 Pa. $4.95

PIANO TUNING, J. Cree Fischer. Clearest, best book for beginner, amateur. Simple repairs, raising dropped notes, tuning by easy method of flattened fifths. No previous skills needed. 4 illustrations. 201pp. 5⅜ x 8½. 23267-0 Pa. $6.95

A SOURCE BOOK IN THEATRICAL HISTORY, A. M. Nagler. Contemporary observers on acting, directing, make-up, costuming, stage props, machinery, scene design, from Ancient Greece to Chekhov. 611pp. 5⅜ x 8½. 20515-0 Pa. $12.95

THE COMPLETE NONSENSE OF EDWARD LEAR, Edward Lear. All nonsense limericks, zany alphabets, Owl and Pussycat, songs, nonsense botany, etc., illustrated by Lear. Total of 320pp. 5⅜ x 8½. (USO) 20167-8 Pa. $6.95

VICTORIAN PARLOUR POETRY: An Annotated Anthology, Michael R. Turner. 117 gems by Longfellow, Tennyson, Browning, many lesser-known poets. "The Village Blacksmith," "Curfew Must Not Ring Tonight," "Only a Baby Small," dozens more, often difficult to find elsewhere. Index of poets, titles, first lines. xxiii + 325pp. 5⅜ x 8¼. 27044-0 Pa. $8.95

DUBLINERS, James Joyce. Fifteen stories offer vivid, tightly focused observations of the lives of Dublin's poorer classes. At least one, "The Dead," is considered a masterpiece. Reprinted complete and unabridged from standard edition. 160pp. 5³⁄₁₆ x 8¼. 26870-5 Pa. $1.00

THE HAUNTED MONASTERY and THE CHINESE MAZE MURDERS, Robert van Gulik. Two full novels by van Gulik, set in 7th-century China, continue adventures of Judge Dee and his companions. An evil Taoist monastery, seemingly supernatural events; overgrown topiary maze hides strange crimes. 27 illustrations. 328pp. 5⅜ x 8½. 23502-5 Pa. $8.95

THE BOOK OF THE SACRED MAGIC OF ABRAMELIN THE MAGE, translated by S. MacGregor Mathers. Medieval manuscript of ceremonial magic. Basic document in Aleister Crowley, Golden Dawn groups. 268pp. 5⅜ x 8½. 23211-5 Pa. $8.95

NEW RUSSIAN-ENGLISH AND ENGLISH-RUSSIAN DICTIONARY, M. A. O'Brien. This is a remarkably handy Russian dictionary, containing a surprising amount of information, including over 70,000 entries. 366pp. 4½ x 6⅛. 20208-9 Pa. $9.95

HISTORIC HOMES OF THE AMERICAN PRESIDENTS, Second, Revised Edition, Irvin Haas. A traveler's guide to American Presidential homes, most open to the public, depicting and describing homes occupied by every American President from George Washington to George Bush. With visiting hours, admission charges, travel routes. 175 photographs. Index. 160pp. 8¼ x 11. 26751-2 Pa. $11.95

NEW YORK IN THE FORTIES, Andreas Feininger. 162 brilliant photographs by the well-known photographer, formerly with *Life* magazine. Commuters, shoppers, Times Square at night, much else from city at its peak. Captions by John von Hartz. 181pp. 9¼ x 10¾. 23585-8 Pa. $12.95

INDIAN SIGN LANGUAGE, William Tomkins. Over 525 signs developed by Sioux and other tribes. Written instructions and diagrams. Also 290 pictographs. 111pp. 6⅛ x 9¼. 22029-X Pa. $3.95

ANATOMY: A Complete Guide for Artists, Joseph Sheppard. A master of figure drawing shows artists how to render human anatomy convincingly. Over 460 illustrations. 224pp. 8⅜ x 11¼. 27279-6 Pa. $10.95

MEDIEVAL CALLIGRAPHY: Its History and Technique, Marc Drogin. Spirited history, comprehensive instruction manual covers 13 styles (ca. 4th century thru 15th). Excellent photographs; directions for duplicating medieval techniques with modern tools. 224pp. 8⅜ x 11¼. 26142-5 Pa. $12.95

DRIED FLOWERS: How to Prepare Them, Sarah Whitlock and Martha Rankin. Complete instructions on how to use silica gel, meal and borax, perlite aggregate, sand and borax, glycerine and water to create attractive permanent flower arrangements. 12 illustrations. 32pp. 5⅜ x 8½. 21802-3 Pa. $1.00

EASY-TO-MAKE BIRD FEEDERS FOR WOODWORKERS, Scott D. Campbell. Detailed, simple-to-use guide for designing, constructing, caring for and using feeders. Text, illustrations for 12 classic and contemporary designs. 96pp. 5⅜ x 8½. 25847-5 Pa. $2.95

SCOTTISH WONDER TALES FROM MYTH AND LEGEND, Donald A. Mackenzie. 16 lively tales tell of giants rumbling down mountainsides, of a magic wand that turns stone pillars into warriors, of gods and goddesses, evil hags, powerful forces and more. 240pp. 5⅜ x 8½. 29677-6 Pa. $6.95

THE HISTORY OF UNDERCLOTHES, C. Willett Cunnington and Phyllis Cunnington. Fascinating, well-documented survey covering six centuries of English undergarments, enhanced with over 100 illustrations: 12th-century laced-up bodice, footed long drawers (1795), 19th-century bustles, 19th-century corsets for men, Victorian "bust improvers," much more. 272pp. 5⅜ x 8¼. 27124-2 Pa. $9.95

ARTS AND CRAFTS FURNITURE: The Complete Brooks Catalog of 1912, Brooks Manufacturing Co. Photos and detailed descriptions of more than 150 now very collectible furniture designs from the Arts and Crafts movement depict davenports, settees, buffets, desks, tables, chairs, bedsteads, dressers and more, all built of solid, quarter-sawed oak. Invaluable for students and enthusiasts of antiques, Americana and the decorative arts. 80pp. 6½ x 9¼. 27471-3 Pa. $8.95

HOW WE INVENTED THE AIRPLANE: An Illustrated History, Orville Wright. Fascinating firsthand account covers early experiments, construction of planes and motors, first flights, much more. Introduction and commentary by Fred C. Kelly. 76 photographs. 96pp. 8¼ x 11. 25662-6 Pa. $8.95

THE ARTS OF THE SAILOR: Knotting, Splicing and Ropework, Hervey Garrett Smith. Indispensable shipboard reference covers tools, basic knots and useful hitches; handsewing and canvas work, more. Over 100 illustrations. Delightful reading for sea lovers. 256pp. 5⅜ x 8½. 26440-8 Pa. $7.95

FRANK LLOYD WRIGHT'S FALLINGWATER: The House and Its History, Second, Revised Edition, Donald Hoffmann. A total revision—both in text and illustrations—of the standard document on Fallingwater, the boldest, most personal architectural statement of Wright's mature years, updated with valuable new material from the recently opened Frank Lloyd Wright Archives. "Fascinating"—*The New York Times.* 116 illustrations. 128pp. 9¼ x 10¾. 27430-6 Pa. $11.95

PHOTOGRAPHIC SKETCHBOOK OF THE CIVIL WAR, Alexander Gardner. 100 photos taken on field during the Civil War. Famous shots of Manassas Harper's Ferry, Lincoln, Richmond, slave pens, etc. 244pp. 10⅝ x 8¼.   22731-6 Pa. $9.95

FIVE ACRES AND INDEPENDENCE, Maurice G. Kains. Great back-to-the-land classic explains basics of self-sufficient farming. The one book to get. 95 illustrations. 397pp. 5⅜ x 8½.   20974-1 Pa. $7.95

SONGS OF EASTERN BIRDS, Dr. Donald J. Borror. Songs and calls of 60 species most common to eastern U.S.: warblers, woodpeckers, flycatchers, thrushes, larks, many more in high-quality recording.   Cassette and manual 99912-2 $9.95

A MODERN HERBAL, Margaret Grieve. Much the fullest, most exact, most useful compilation of herbal material. Gigantic alphabetical encyclopedia, from aconite to zedoary, gives botanical information, medical properties, folklore, economic uses, much else. Indispensable to serious reader. 161 illustrations. 888pp. 6½ x 9¼. 2-vol. set. (USO)   Vol. I: 22798-7 Pa. $9.95
    Vol. II: 22799-5 Pa. $9.95

HIDDEN TREASURE MAZE BOOK, Dave Phillips. Solve 34 challenging mazes accompanied by heroic tales of adventure. Evil dragons, people-eating plants, blood-thirsty giants, many more dangerous adversaries lurk at every twist and turn. 34 mazes, stories, solutions. 48pp. 8¼ x 11.   24566-7 Pa. $2.95

LETTERS OF W. A. MOZART, Wolfgang A. Mozart. Remarkable letters show bawdy wit, humor, imagination, musical insights, contemporary musical world; includes some letters from Leopold Mozart. 276pp. 5⅜ x 8½.   22859-2 Pa. $7.95

BASIC PRINCIPLES OF CLASSICAL BALLET, Agrippina Vaganova. Great Russian theoretician, teacher explains methods for teaching classical ballet. 118 illus-trations. 175pp. 5⅜ x 8½.   22036-2 Pa. $5.95

THE JUMPING FROG, Mark Twain. Revenge edition. The original story of The Celebrated Jumping Frog of Calaveras County, a hapless French translation, and Twain's hilarious "retranslation" from the French. 12 illustrations. 66pp. 5⅜ x 8½.
    22686-7 Pa. $3.95

BEST REMEMBERED POEMS, Martin Gardner (ed.). The 126 poems in this superb collection of 19th- and 20th-century British and American verse range from Shelley's "To a Skylark" to the impassioned "Renascence" of Edna St. Vincent Millay and to Edward Lear's whimsical "The Owl and the Pussycat." 224pp. 5⅜ x 8½.
    27165-X Pa. $4.95

COMPLETE SONNETS, William Shakespeare. Over 150 exquisite poems deal with love, friendship, the tyranny of time, beauty's evanescence, death and other themes in language of remarkable power, precision and beauty. Glossary of archaic terms. 80pp. 5³⁄₁₆ x 8¼.   26686-9 Pa. $1.00

BODIES IN A BOOKSHOP, R. T. Campbell. Challenging mystery of blackmail and murder with ingenious plot and superbly drawn characters. In the best tradition of British suspense fiction. 192pp. 5⅜ x 8½.   24720-1 Pa. $6.95

THE WIT AND HUMOR OF OSCAR WILDE, Alvin Redman (ed.). More than 1,000 ripostes, paradoxes, wisecracks: Work is the curse of the drinking classes; I can resist everything except temptation; etc. 258pp. 5⅜ x 8½. 20602-5 Pa. $5.95

SHAKESPEARE LEXICON AND QUOTATION DICTIONARY, Alexander Schmidt. Full definitions, locations, shades of meaning in every word in plays and poems. More than 50,000 exact quotations. 1,485pp. 6½ x 9¼. 2-vol. set.
Vol. 1: 22726-X Pa. $16.95
Vol. 2: 22727-8 Pa. $16.95

SELECTED POEMS, Emily Dickinson. Over 100 best-known, best-loved poems by one of America's foremost poets, reprinted from authoritative early editions. No comparable edition at this price. Index of first lines. 64pp. 5$\frac{3}{16}$ x 8¼.
26466-1 Pa. $1.00

CELEBRATED CASES OF JUDGE DEE (DEE GOONG AN), translated by Robert van Gulik. Authentic 18th-century Chinese detective novel; Dee and associates solve three interlocked cases. Led to van Gulik's own stories with same characters. Extensive introduction. 9 illustrations. 237pp. 5⅜ x 8½. 23337-5 Pa. $6.95

THE MALLEUS MALEFICARUM OF KRAMER AND SPRENGER, translated by Montague Summers. Full text of most important witchhunter's "bible," used by both Catholics and Protestants. 278pp. 6⅝ x 10. 22802-9 Pa. $12.95

SPANISH STORIES/CUENTOS ESPAÑOLES. A Dual-Language Book, Angel Flores (ed.). Unique format offers 13 great stories in Spanish by Cervantes, Borges, others. Faithful English translations on facing pages. 352pp. 5⅜ x 8½.
25399-6 Pa. $8.95

THE CHICAGO WORLD'S FAIR OF 1893: A Photographic Record, Stanley Appelbaum (ed.). 128 rare photos show 200 buildings, Beaux-Arts architecture, Midway, original Ferris Wheel, Edison's kinetoscope, more. Architectural emphasis; full text. 116pp. 8¼ x 11. 23990-X Pa. $9.95

OLD QUEENS, N.Y., IN EARLY PHOTOGRAPHS, Vincent F. Seyfried and William Asadorian. Over 160 rare photographs of Maspeth, Jamaica, Jackson Heights, and other areas. Vintage views of DeWitt Clinton mansion, 1939 World's Fair and more. Captions. 192pp. 8⅞ x 11. 26358-4 Pa. $12.95

CAPTURED BY THE INDIANS: 15 Firsthand Accounts, 1750-1870, Frederick Drimmer. Astounding true historical accounts of grisly torture, bloody conflicts, relentless pursuits, miraculous escapes and more, by people who lived to tell the tale. 384pp. 5⅜ x 8½. 24901-8 Pa. $8.95

THE WORLD'S GREAT SPEECHES, Lewis Copeland and Lawrence W. Lamm (eds.). Vast collection of 278 speeches of Greeks to 1970. Powerful and effective models; unique look at history. 842pp. 5⅜ x 8½. 20468-5 Pa. $14.95

THE BOOK OF THE SWORD, Sir Richard F. Burton. Great Victorian scholar/adventurer's eloquent, erudite history of the "queen of weapons"–from prehistory to early Roman Empire. Evolution and development of early swords, variations (sabre, broadsword, cutlass, scimitar, etc.), much more. 336pp. 6⅛ x 9¼.
25434-8 Pa. $9.95

AUTOBIOGRAPHY: The Story of My Experiments with Truth, Mohandas K. Gandhi. Boyhood, legal studies, purification, the growth of the Satyagraha (nonviolent protest) movement. Critical, inspiring work of the man responsible for the freedom of India. 480pp. 5⅜ x 8½. (USO)   24593-4 Pa. $8.95

CELTIC MYTHS AND LEGENDS, T. W. Rolleston. Masterful retelling of Irish and Welsh stories and tales. Cuchulain, King Arthur, Deirdre, the Grail, many more. First paperback edition. 58 full-page illustrations. 512pp. 5⅜ x 8½.   26507-2 Pa. $9.95

THE PRINCIPLES OF PSYCHOLOGY, William James. Famous long course complete, unabridged. Stream of thought, time perception, memory, experimental methods; great work decades ahead of its time. 94 figures. 1,391pp. 5⅜ x 8½. 2-vol. set.
Vol. I: 20381-6 Pa. $12.95
Vol. II: 20382-4 Pa. $12.95

THE WORLD AS WILL AND REPRESENTATION, Arthur Schopenhauer. Definitive English translation of Schopenhauer's life work, correcting more than 1,000 errors, omissions in earlier translations. Translated by E. F. J. Payne. Total of 1,269pp. 5⅜ x 8½. 2-vol. set.   Vol. 1: 21761-2 Pa. $11.95
Vol. 2: 21762-0 Pa. $12.95

MAGIC AND MYSTERY IN TIBET, Madame Alexandra David-Neel. Experiences among lamas, magicians, sages, sorcerers, Bonpa wizards. A true psychic discovery. 32 illustrations. 321pp. 5⅜ x 8½. (USO)   22682-4 Pa. $8.95

THE EGYPTIAN BOOK OF THE DEAD, E. A. Wallis Budge. Complete reproduction of Ani's papyrus, finest ever found. Full hieroglyphic text, interlinear transliteration, word-for-word translation, smooth translation. 533pp. 6½ x 9¼.
21866-X Pa. $10.95

MATHEMATICS FOR THE NONMATHEMATICIAN, Morris Kline. Detailed, college-level treatment of mathematics in cultural and historical context, with numerous exercises. Recommended Reading Lists. Tables. Numerous figures. 641pp. 5⅜ x 8½.
24823-2 Pa. $11.95

THEORY OF WING SECTIONS: Including a Summary of Airfoil Data, Ira H. Abbott and A. E. von Doenhoff. Concise compilation of subsonic aerodynamic characteristics of NACA wing sections, plus description of theory. 350pp. of tables. 693pp. 5⅜ x 8½.   60586-8 Pa. $14.95

THE RIME OF THE ANCIENT MARINER, Gustave Doré, S. T. Coleridge. Doré's finest work; 34 plates capture moods, subtleties of poem. Flawless full-size reproductions printed on facing pages with authoritative text of poem. "Beautiful. Simply beautiful."—*Publisher's Weekly.* 77pp. 9¼ x 12.   22305-1 Pa. $6.95

NORTH AMERICAN INDIAN DESIGNS FOR ARTISTS AND CRAFTSPEOPLE, Eva Wilson. Over 360 authentic copyright-free designs adapted from Navajo blankets, Hopi pottery, Sioux buffalo hides, more. Geometrics, symbolic figures, plant and animal motifs, etc. 128pp. 8⅜ x 11. (EUK)   25341-4 Pa. $8.95

SCULPTURE: Principles and Practice, Louis Slobodkin. Step-by-step approach to clay, plaster, metals, stone; classical and modern. 253 drawings, photos. 255pp. 8⅛ x 11.
22960-2 Pa. $11.95

THE INFLUENCE OF SEA POWER UPON HISTORY, 1660–1783, A. T. Mahan. Influential classic of naval history and tactics still used as text in war colleges. First paperback edition. 4 maps. 24 battle plans. 640pp. 5⅜ x 8½. 25509-3 Pa. $12.95

THE STORY OF THE TITANIC AS TOLD BY ITS SURVIVORS, Jack Winocour (ed.). What it was really like. Panic, despair, shocking inefficiency, and a little heroism. More thrilling than any fictional account. 26 illustrations. 320pp. 5⅜ x 8½. 20610-6 Pa. $8.95

FAIRY AND FOLK TALES OF THE IRISH PEASANTRY, William Butler Yeats (ed.). Treasury of 64 tales from the twilight world of Celtic myth and legend: "The Soul Cages," "The Kildare Pooka," "King O'Toole and his Goose," many more. Introduction and Notes by W. B. Yeats. 352pp. 5⅜ x 8½. 26941-8 Pa. $8.95

BUDDHIST MAHAYANA TEXTS, E. B. Cowell and Others (eds.). Superb, accurate translations of basic documents in Mahayana Buddhism, highly important in history of religions. The Buddha-karita of Asvaghosha, Larger Sukhavativyuha, more. 448pp. 5⅜ x 8½. 25552-2 Pa. $12.95

ONE TWO THREE . . . INFINITY: Facts and Speculations of Science, George Gamow. Great physicist's fascinating, readable overview of contemporary science: number theory, relativity, fourth dimension, entropy, genes, atomic structure, much more. 128 illustrations. Index. 352pp. 5⅜ x 8½. 25664-2 Pa. $8.95

ENGINEERING IN HISTORY, Richard Shelton Kirby, et al. Broad, nontechnical survey of history's major technological advances: birth of Greek science, industrial revolution, electricity and applied science, 20th-century automation, much more. 181 illustrations. ". . . excellent . . ."–Isis. Bibliography. vii + 530pp. 5⅜ x 8¼. 26412-2 Pa. $14.95

DALÍ ON MODERN ART: The Cuckolds of Antiquated Modern Art, Salvador Dalí. Influential painter skewers modern art and its practitioners. Outrageous evaluations of Picasso, Cézanne, Turner, more. 15 renderings of paintings discussed. 44 calligraphic decorations by Dalí. 96pp. 5⅜ x 8½. (USO) 29220-7 Pa. $4.95

ANTIQUE PLAYING CARDS: A Pictorial History, Henry René D'Allemagne. Over 900 elaborate, decorative images from rare playing cards (14th–20th centuries): Bacchus, death, dancing dogs, hunting scenes, royal coats of arms, players cheating, much more. 96pp. 9¼ x 12¼. 29265-7 Pa. $11.95

MAKING FURNITURE MASTERPIECES: 30 Projects with Measured Drawings, Franklin H. Gottshall. Step-by-step instructions, illustrations for constructing handsome, useful pieces, among them a Sheraton desk, Chippendale chair, Spanish desk, Queen Anne table and a William and Mary dressing mirror. 224pp. 8⅛ x 11¼. 29338-6 Pa. $13.95

THE FOSSIL BOOK: A Record of Prehistoric Life, Patricia V. Rich et al. Profusely illustrated definitive guide covers everything from single-celled organisms and dinosaurs to birds and mammals and the interplay between climate and man. Over 1,500 illustrations. 760pp. 7½ x 10⅛. 29371-8 Pa. $29.95

Prices subject to change without notice.

Available at your book dealer or write for free catalog to Dept. GI, Dover Publications, Inc., 31 East 2nd St., Mineola, N.Y. 11501. Dover publishes more than 500 books each year on science, elementary and advanced mathematics, biology, music, art, literary history, social sciences and other areas.